Superconductivity
Experimenting
In a New
Technology

ADVANCED TECHNOLOGY SERIES

Other Books in the Advanced Technology Series

Experiments with EPROMs

by Dave Prochnow
Edited by Lisa A. Doyle

Experiments in Artificial Neural Networks

by Ed Rietman
Edited by David Gauthier

CMOS Technology

by Dave Prochnow and D.J. Branning
Edited by David Gauthier

Experiments in Gallium Arsenide Technology

by D.J. Branning and Dave Prochnow
Edited by Lisa A. Doyle

SUPERCONDUCTIVITY EXPERIMENTING IN A NEW TECHNOLOGY

Dave Prochnow

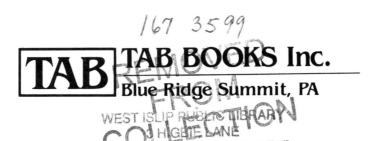

167 3599

TAB BOOKS Inc.

Blue Ridge Summit, PA

FIRST EDITION
FIRST PRINTING

Copyright © 1989 by TAB BOOKS Inc.
Printed in the United States of America.

Library of Congress Cataloging in Publication Data

Prochnow, Dave.
    Superconductivity : experiments in a new technology / by Dave Prochnow.
        p.   cm.
    Bibliography: p.
    Includes index.
    ISBN 0-8306-1432-X   ISBN 0-8306-3132-1 (pbk.)
    1. Superconductivity. 2. Superconductors. I. Title.
QC611.92.P76 1988
537.6'23—dc19   88-22436
                                          CIP

TAB BOOKS Inc. offers software for sale. For information and a catalog, please contact TAB Software Department, Blue Ridge Summit, PA 17294-0850.

Questions regarding the content of this book should be addressed to:

Reader Inquiry Branch
TAB BOOKS Inc.
Blue Ridge Summit, PA 17294-0214

Front cover photograph courtesy of AT&T Bell Laboratories.

To my best friend and wife, Kathy

Warning and Disclaimer

This book deals with subjects, materials, and procedures that can be hazardous to your health. Use extreme caution when performing this book's experiments. Do not attempt to perform any of these experiments unless you fully comprehend all of the materials' associated handling precautions. If you lack this information, consult with a university-level chemistry or physics instructor.

Although every possible safeguard has been employed in ensuring the accuracy of this book's information, neither the author nor TAB BOOKS, Inc. can be held liable for damages or injuries that could result from the application, misinterpretation, and/or misapplication of the materials and procedures that are discussed in this book.

Contents

Acknowledgments

Two valuable contributions were made by Colorado Superconductor, Inc. and Furuuchi Chemical Corporation during the preparation of this book. These informative demonstration kits and their accompanying documentation served as vital references for developing this text and its associated projects.

Introduction—
Superconductivity Today

No other technology has made as much revolutionary progress in such a short span of time as the superconductor industry. What began as an unusual phenomenon in 1911 laid dormant for some seventy years, buried inside obscure scientific journals. Then in the late 1980s, a tidal wave of technological advances turned superconductivity into a highly coveted buzzword. Overly exaggerated promises, however, spoiled the rapid growth of superconductivity research into practical applications. Previously labeled the "salvation of society," superconductivity is still at a crossroads between remaining a scientific curiosity and becoming an everyday cost-effective realization.

Today's press is filled with impressive accounts of superconductivity fulfilling this latter preference. For example, high-temperature superconductor research is being slanted toward the rapid development of superconducting thin films and integrated circuits (ICs). One research team at TRW is attempting to design Josephson junctions from superconducting thin films as a step toward manufacturing the first superconducting IC. Although these ICs are designed to operate at a critical temperature of 92 °K, the largest drawback to production is the actual structure of the Josephson junction. Normal input/output (I/O) operations with this type of circuit are limited by its inherent two-terminal structure. Therefore, TRW's research team must also develop a better

Josephson junction "mousetrap" before a superconductor IC will be practical.

Even the once balmy critical temperature of 92°K is being threatened by a research group at the Georgia Institute of Technology. These physicists have tested superconductive ceramics with critical temperatures near 500°K. If these materials can be successively isolated and manufactured (these are the two major obstacles currently hampering the Georgia Tech team), then room-temperature superconductors will be possible.

Supplementing this theoretical temperature elevation discovery, IBM Corporation's Almaden Research Center has a confirmed 125°K ceramic that offers viable promise to the goal of a room-temperature superconductor (see Fig. A). The IBM research team's high-temperature ceramic is an oxide compound of thalium, barium, calcium, and copper.

At the University of Arkansas at Fayetteville, another group of scientists developed a different ceramic based on these same four elements that became superconductive at the critical temperature of 106°K. Interestingly enough, IBM's confidence in their ceramic is best voiced by their application for patent protection in March of 1988.

On other design fronts, scientists at General Motors Research Laboratories are making significant contributions toward the fabrication of fast and inexpensive superconducting thin films. GM's unique approach to thin film compositions uses metallo-organic deposition techniques. This simple, low-cost process is able to quickly produce yttrium barium copper oxide superconducting thin films on strontium titanate and barium titanate substrates that operate with a critical temperature of 90°K.

An alternate thin film production technique is employed by Bell Communications Research (Bellcore). This process uses a pulsed excimer laser as an ion-beam sputter-coating device (see Fig. B). In operation, each pulse from the laser places a fixed coating of superconductive material on a substrate. Following a predetermined number of pulsations (usually 5,000 pulses for a 2- to 3-micron thin film), the completed superconducting thin film is ready for application.

One other superconductor production effort is in the quest for a mass-produced superconductive ceramic. This development will signal the direct application of superconductors within magnets and electrical-field-generation equipment. A group of scientists at the University of California, Davis has devised a method for turning the yttrium barium copper oxide ceramic powder into a finished production-grade superconducting cylinder. The entire process takes under 10 seconds, which helps to retain the superconductor's high critical temperature.

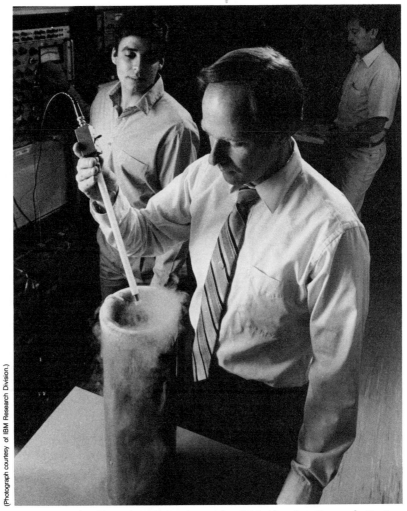

Fig. A. A scientific research group at the IBM Almaden Research Center synthesizes high-temperature superconductors in bulk quantities.

As far as actual applications for these superconductors and superconducting thin films are concerned, there are several exciting research programs currently realizing success. Westinghouse Electric Corporation and the National Bureau of Standards are currently exploring a means for designing low-resistance electrical contacts on high-temperature superconductors. This research is a significant step in minimizing the destructive effects that resistance heat has on the superconducting state.

One final example of current developments in superconductivity includes an optically driven superconducting switch designed by Hitachi, Ltd. Operating in the high-temperature region of 88°K, this switch is

Fig. B. GE Research and Development Center scientists apply superconducting films to silicon substrates through electron beam evaporation.

composed of two thin-film electrodes that are separated by an ultra-thin film superconducting trench. This trench acts as a Cooper pair "highway" for the free flow of electrons between the electrodes. When this trench area is exposed to light, however, the superconductive state is eliminated. Therefore, this switch is controlled through the normal and superconducting states of the electron trench. This switching action is similar to the "off" and "on" states found in conventional switches.

By and large, this sudden proliferation in superconductivity research owes a great deal of its success to a mammoth project disclosed by President Ronald Reagan in July 1987 (see Fig. C). This 11-point Superconductivity Initiative was detailed at the Federal Conference on Commercial Applications of Superconductors. Increased patent protection, antitrust law legislation, the establishment of regional superconductivity centers, and $150 million in research and development (R&D) for Department of Defense (DOD) projects were four of the more important points that were addressed by President Reagan's Superconductivity Initiative.

As is clearly evident from these technological highlights, the current state of superconductivity is one of research. As such, this technology is thirsting for knowledge. One way of providing this knowledge is through experimentation. Unfortunately, experimentation with a technology as complex as superconductivity requires extensive background information. Both of these factors have been major stumbling blocks for rapid advances in superconductivity. Attempting to squelch both of these limiting factors, this book provides an advanced

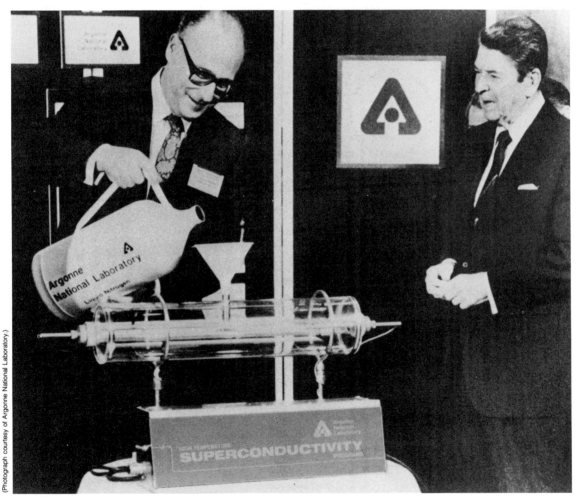

(Photograph courtesy of Argonne National Laboratory.)

Fig. C. President Ronald Reagan receives a lesson in superconductivity from Alan Schriesheim, the director of Argonne National Laboratory (left).

examination of superconductivity supplemented with simple demonstration experiments. In short, this book is a definitive course on superconductivity.

One drawback to providing extensive coverage of a complex scientific topic like superconductivity is that the depth of generated information exceeds the knowledge of the average experimenter. In an attempt to make superconductivity accessible to all readers, every effort has been made to thoroughly cover all aspects of superconductivity in a progressive manner, beginning with elementary generalizations and advancing to theoretical topics. Be forewarned, however, that this is not an elementary treatment of superconductivity for the layperson. There is no handholding of the reader through the basics of inorganic chemistry, elementary thermodynamics, and introductory quantum mechanics. If these topics are alien to you, then you should obtain introductory texts prior to reading this book.

There are numerous formulae and format conventions used throughout this book. In the case of equation and expression development, the following units should be understood by the reader:

Blotzmann's constant = k = 1.380622×10^{-16} erg/°K

Planck's constant = h = 6.626196×10^{-27} erg-sec

Speed of light in a vacuum = c = 2.9979250×10^{10} cm/sec

Electron charge = e = 1.602×10^{-19} coulomb

Electron rest mass = m_e = 9.109558×10^{-28} g

Proton rest mass = m_p = 1.672614×10^{-24} g

Flux = Wb = 10^8 lines

Flux density = T = 1 Wbm^{-2}

Field strength = Am^{-1} = 0.01259 Oe

Furthermore, the text itself contains hundreds of literature and reference citations that are not footnoted. All of this material is derived from the sources cited in the bibliography of this book. If you need additional support documentation regarding a text topic, consult these reference books for supplementary information.

What is the next discovery threshold for superconductivity? Where will the critical temperature race end? Will superconductors ever be realized in real-world applications? Maybe your research, based on the information you gain from this book, will form the next major event in the time line of superconductor research.

1

An Introduction to Superconductivity

If cinema is a reliable prognosticator of reality, then the near future will realize countless significant discoveries in medical, military, and recreational disciplines. The technology professed to provide the impetus for these remarkable breakthroughs is summarized with one word—*superconductivity*. Motion pictures such as *Brainstorm* dramatize the effects of superconductivity as the last "lock," which when opened, can achieve a "total mental array processing" device. In this case, fiction has portrayed superconductivity as being able to unlock the secrets of the human mind. Is this science fiction tomorrow's science fact? Probably not, but the real-world applications that will be factual derivatives from superconductivity might be no less startling in their practical significance.

By considering superconductivity to be a lock on a future revolution in science, gaining access to these scientific riches will require the application of some strange keys. Barium, copper oxide, lanthanum, niobium, and yttrium form the keys to the solution of superconductivity. Unfortunately, turning the tumblers of the superconductivity lock is not solved by a simple, rare, earth compound. There are six factors that must be carefully monitored in pursuit of achieving superconductivity: temperature, resistance, critical current, magnetic field, oxygen site, and stability.

Temperature. Of these six superconductivity performance criteria, the requirement for frigid operating temperatures is the most frequently voiced. Conversely, this temperature requirement is the one area in superconductivity that is exhibiting the greatest degree of change.

Recent advances in the selection and preparation of superconductors has elevated the operating temperature at which materials can superconduct from 23 degrees Kelvin to a blistering 294°K. These colder temperatures were obtained through superconducting magnets constructed from niobium in 1973. The first significant advance in superconductivity was reached in 1986 by a team of scientists at the IBM Research Laboratory in Zurich, Switzerland. This team, headed by Johannes Georg Bednorz and Karl Alex Müller, increased the operating temperature for superconductors by 12 degrees to 35°K (see Fig. 1-1). Their warm temperature superconductor was formed from a barium, lanthanum, and copper oxide compound. High-temperature super-conductivity was becoming a reality.

In 1987, the last major hurdle in obtaining manageable super-conductor operating temperatures was surmounted. Two research teams located at the University of Houston and the University of Alabama at

(Photograph courtesy of IBM Research Division.)

*Fig. 1-1. J. Georg Bednorz and K. Alex Müller
at IBM's Zurich Research Laboratory.*

Fig. 1-2. Dr. Paul Chu, T.L.L. Temple Chair in Science, University of Houston.

Huntsville co-developed a compound from copper oxide and yttrium that superconducted at 98°K (see Fig. 1-2). This temperature increase marked the advent of reliable, reproducible, high-temperature superconductors.

Following on the heels of this discovery, other research teams have reported experimental results with compounds that superconduct at 294°K. Unfortunately, these superconductors are only present in controlled experimental environments and lack the reliable and reproducible qualities of the 98°K high-temperature superconductors.

Resistance. A strict guideline that is applied to any material that purports to be a superconductor is the absence of direct current resistance. In this criterion, a zero resistivity must be maintained by the superconductor until it reaches its critical current (see Fig. 1-3). Not only is resistance current sensitive, but it is also variable over a fixed temperature range. For example, an increase in temperature of 100°K to 198°K with yttrium copper oxides will remove the complete resistivity of a superconductor.

Fig. 1-3. AT&T Bell Laboratories metallurgist, Sungho Jin, tests the resistivity of a melt-textured growth bulk superconductor.

Critical Current. Even under optimal operating temperatures, a superconductor loses its total resistivity at a specific current level called the *critical current*. Shadowing research into elevation of operating temperatures, a group of scientists at the IBM Research Laboratory have designed superconducting thin films that exhibit a critical current of 100,000 A/cm^2. Superconductors with a high critical current are necessary for turning laboratory experiments into production microelectronics.

Magnetic Field. In addition to displaying a complete absence of current resistance, superconductors also generate a magnetic field. This magnetic field generation is called the *Meissner effect*. One commercial application of the Meissner effect is to float a magnet over a superconductor. This "floating magnet" application is erroneously perceived as a product of superconductivity, whereas the levitated magnet is simply providing a visual demonstration of the formation of a magnetic field (see Fig. 1-4).

The final two factors used in the formation and evaluation of superconductivity are presently at a theoretical level of interpretation.

Oxygen Site. Both oxygen content and oxygen's atomic arrangement hold roles in the presence or absence of superconductivity from a given compound. In a typical triple perovskite superconductor, there are four oxygen sites that contribute to the compound's oxygen content.

Stability. Superconductors must exhibit stability when reacting to other compounds. Even the reaction of a superconductor to water and air is an unpredictable factor in its function.

(Photograph courtesy of Argonne National Laboratory.)

Fig. 1-4. Magnetic levitation via the Meissner effect.

THE MAKING OF A SUPERCONDUCTOR

Turning superconductivity theory into a superconducting example is readily obtainable from high-temperature superconductors. There are two separate phases to superconductor construction: preparation and cooling.

Ceramic Preparation. A successful superconducting preparation must blend a rare earth element like yttrium (even though yttrium is NOT a true rare earth element, it is frequently included in this lanthanide series of elements) with a strong oxygen supplier like copper oxide (see Fig. 1-5). This combination process is achieved through a series of grinding, heating, and cooling steps. Depending on the final molecular structure of this superconductor, the exact sequence and time/temperature specifications vary. Frequently, the result from this combination process is a superconducting powder that must be formed into either a disk or a pellet shape prior to any conventional application usage.

The formation of a superconducting disk is easily achieved through a press. A typical press capable of producing 10,000 to 20,000 psi is required for compacting the black superconducting powder into a final disk shape.

Cooling. Once the disk has been prepared, the material must be cooled in order to become superconductive. One common means of producing the required temperature reduction is through the immersion of the disk in a bath of liquid helium. Although liquid helium is a superb coolant ($4.2\,^\circ K$) its cost and sensitive handling requirements make the use of liquid nitrogen far superior. Liquid nitrogen remains stable at a balmy $77\,^\circ K$. Therefore, the elevated temperature of liquid nitrogen is only applicable to high-temperature superconductors.

SUPERCONDUCTORS AT WORK

Superconductivity, in itself, is open to hybrid applications where the benefits of high-temperature superconductors can be directly interfaced with present technology. Two areas where hybrid superconductor application's are most practical are in microelectronics and superconducting quantum interference devices (SQUIDs).

Microelectronics. In joining contemporary semiconductor technologies with superconductors, the absence of current resistivity enables the microelectronic device to operate more efficiently at lower temperatures. This temperature savings is most evident when

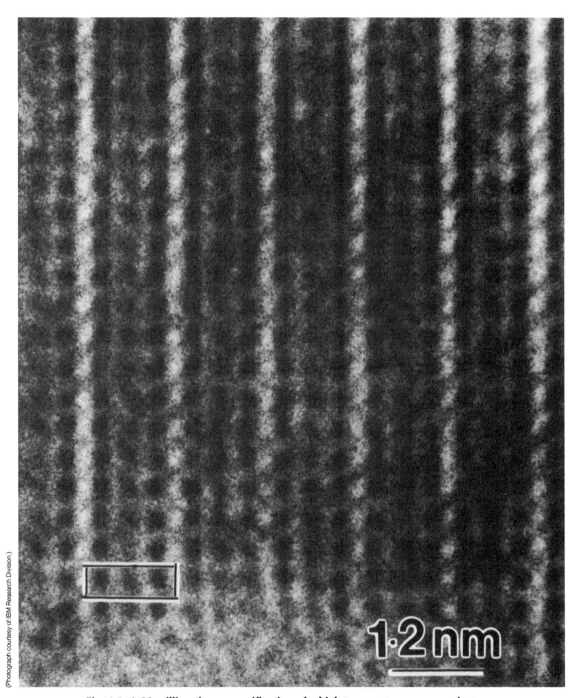

(Photograph courtesy of IBM Research Division.)

Fig. 1-5. A 20-million times magnification of a high-temperature superconductor.

superconductors are applied to power buses and device intraconnection traces.

SQUIDs. Josephson junctions are superconducting switches that are similar in operation to transistors, but, they operate at frequencies 100 times faster than conventional transistors. One application for Josephson junctions is in the connection of two junctions together for detecting minute magnetic fields. These sensitive magnetic field detectors are known as *superconducting quantum interference devices* or *SQUIDs* (see Fig. 1-6).

Interference devices like SQUIDs are used for monitoring magnetic fields. This monitoring is especially important in biomedical magnetism and magnetic resonance imaging studies, as well as gravitational wave exploration. Currently, low-temperature SQUIDs are used for these magnetic and gravitational studies. By replacing these SQUIDs with high-temperature SQUIDs, a liquid nitrogen cooling system can be substituted for the current liquid helium cooling system. The result from this

(Photograph courtesy of IBM Research Division.)

Fig. 1-6. A 500-times magnification of a high-temperature SQUID.

(Photograph courtesy of Argonne National Laboratory.)

Fig. 1-7. Argonne National Laboratory researcher, Mr. Jim Jorgensen evaluates superconductors with a computer-controlled Special Environment Powder Diffractometer.

substitution will be a reduction in SQUID bulk and a decrease in the cost of building the high-temperature SQUID.

Superconductivity imaging equipment such as *nuclear magnetic resonance* (NMR) or *magnetic resonance imaging* (MRI) devices are capable of generating detailed images of internal structures. An MRI device can generate an image of a patient's heart without the need for X-rays or tracing barium dye injections. These MRI devices work by placing the target structure inside a powerful magnetic field generated by a superconducting electromagnet.

An MRI device exposes a target subject to a strong magnetic field generated by a superconducting electromagnetic coil. When the subject is exposed to this magnetic field, the protons in the target tissue's molecules align themselves relative to the applied magnetic field. An injection of radio frequency energy operated at a correlated resonant frequency excites the protons. When the radio frequency burst decays, the protons return to their former state and release energy. This energy release is detected and used for the creation of an image. By altering the magnetic field, images of internal structures can be represented as anatomic cross-sectional compositions.

There are several unknown variables and design precautions that block the immediate introduction of high-temperature superconductors into these two application areas (see Fig. 1-7).

❖ **Heat**; superconductors require high curing temperatures that can destroy delicate hybrid technologies.

❖ **Chemical contamination**; hybrid technologies use exotic etchants, solvents, and adhesives during their preparation that can react unfavorably with superconductors.

❖ **Effectiveness**; will high-temperature superconductors operate as sensitively as their low-temperature ancestors?

❖ **Heat buildup**; using superconductors in areas of high-temperature production could reduce and eliminate their superconductive benefits.

Providing solutions to these possible pitfalls will help in strengthening superconductivity's position in mainstream technology. Once these problems are overcome, areas for exploitation will include generator design, static magnet construction, and the creation of power and electromechanical devices. At this point, however, superconductivity is advancing at the research level. Understanding the complex elements involved in this research will eventually lead to the translation of science fiction into production fact.

2

Superconductors

Superconductivity technology can trace its origin to the rather recent date of 1908. Heike Kammerlingh Onnes, a Dutch physicist, liquified helium to a temperature below 4.2 °K at this time. Subsequently, Onnes applied his newly found coolant to mercury and measured the electrical resistance through the supercooled element. The results from this experiment indicated that resistivity decreased in the mercury under the influence of liquid helium. Onnes labeled this change in the mercury, at temperatures below 4.15 °K, a superconducting state. Therefore, the initial low-temperature superconductivity experiments were with mercury in liquid helium baths.

Following several months of careful sampling and resistance measurement, Onnes fixed the state of resistivity at 10^{-20} Ωcm. This led Onnes to assume that zero-resistance superconductivity was conclusive in mercury. Other experimenters lent support to this assumption by duplicating Onnes measurements with indium, lead, and tin. Unfortunately, it was discovered that minor magnetic fields of 480 A cm^{-1} disrupted these early elemental superconductors, which shattered the initial hopes of free energy yields (see Fig. 2-1).

The field-dampening nature of these modest magnetic fields virtually eliminated the possibility of superconductive magnet designs.

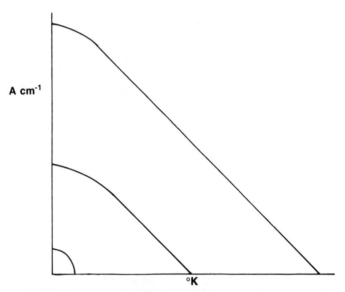

Fig. 2-1. Typical critical field strength curves for temperature.

Additionally, forming the early superconductors into a functional wire were met with equal disappointment. In fact, a study in 1916 proved that superconductive-robbing current was only needed on the surface of a superconducting wire for destroying its inherent zero resistivity.

An interesting footnote to these early attempts at "practical" applications of superconductivity is found in the 1930 construction of a superconducting magnet. This prototype magnet featured a eutectic alloy of bismuth and lead that remained superconductive at a temperature of 4.2 °K and in a magnetic field of 12 kA cm^{-1}. In spite of these promising performance specifications, the current in the alloy remained impractically low and the project was shelved.

Based on these early studies, two important aspects of experimenting with superconductors were determined:

1. **Microscopic theory;** this theory, derived from a 1957 study by John Bardeen, Leon N. Cooper, and J. R. Shrieffer, laid the groundwork for today's superconductor research (this theory is occasionally referred to as the *BCS theory of superconductivity*). In this study, Bardeen, Cooper, and Shrieffer declared an inter-relationship between two opposing forces in superconductivity technology—critical temperature and critical field strength.

2. **Thermodynamics;** a 1929 study showed that there was a thermodynamic relationship between the transformation of the superconductor from its native state into its superconducting state.

In an attempt at providing a thorough understanding into these two significant superconductor aspects, Chapter 3 and Chapter 4 are devoted to the microscopic theory and thermodynamics, respectively. The remainder of this chapter, however, will lay an essential foundation for applying these fundamental superconductor aspects.

ELEMENTARY INORGANIC CHEMISTRY

One of the first definitions that must be made when working with superconductors is the identification of a metal. There are essentially two general classes of elements—metals and nonmetals. A beginning attempt at separating these two groups is based on the ability of elemental oxides to form either bases or acids. In this classification scheme, elements like sodium are considered metals, whereas elements like carbon are labeled nonmetals. Additional characteristics used in the separation of metals and nonmetals include electrical conductivity, luster, and opaqueness.

While these descriptive characteristics are suitable for the general classification of metals and nonmetals, they fail to adequately define the phenomenon of superconductivity in metals. In order to understand the presence of this quality in some metals, it is necessary to explore the atomic structure of metals versus nonmetals.

Essentially, at the atomic level, the difference between metals and nonmetals can be traced to both the number of valence electrons and the number of electron shells. By applying these two qualities to the known elements, the metallic character of an element decreases with the increase in the number of valence electrons. The metallic character increases, however, with the increase in the number of electron shells. The reciprocal of these two statements holds true for the nonmetallic character of an element. In other words, nonmetallic characteristics increase with the increase in the number of valence electrons and the decrease in the number of electron shells.

In reference to the Periodic Table of Elements (see Fig. 2-2), the number of valence electrons increases from left to right and the number of electron shells increases down each column. Therefore, the properties of the elements become increasingly more metallic to the left and down

Fig. 2-2. The Periodic Table of Elements.

IA	IIA	IIIB	IVB	VB	VIB	VIIB	VIII	VIII	VIII	IB	IIB	IIIA	IVA	VA	VIA	VIIA	0
H 1																H 1	He 2
Li 3	Be 4											B 5	C 6	N 7	O 8	F 9	Ne 10
Na 11	Mg 12											Al 13	Si 14	P 15	S 16	Cl 17	Ar 18
K 19	Ca 20	Sc 21	Ti 22	V 23	Cr 24	Mn 25	Fe 26	Co 27	Ni 28	Cu 29	Zn 30	Ga 31	Ge 32	As 34	Se 34	Br 35	Kr 36
Rb 37	Sr 38	Y 39	Zr 40	Nb 41	Mo 42	Tc 43	Ru 44	Rh 45	Pd 46	Ag 47	Cd 48	In 49	Sn 50	Sb 51	Te 52	I 53	Xe 54
Cs 55	Ba 56	*	Hf 72	Ta 73	W 74	Re 75	Os 76	Ir 77	Pt 78	Au 79	Hg 80	Tl 81	Pb 82	Bi 83	Po 84	At 85	Rn 86
Fr 87	Ra 88	**	Unq 104	Unp 105	Unh 106	Uns 107											

*Lanthanide Series

La 57	Ce 58	Pr 59	Nd 60	Pm 61	Sm 62	Eu 63	Gd 64	Tb 65	Dy 66	Ho 67	Er 68	Tm 69	Yb 70	Lu 71

**Actinide Series

Ac 89	Th 90	Pa 91	U 92	Np 93	Pu 94	Am 95	Cm 96	Bk 97	Cf 98	Es 99	Fm 100	Md 101	No 102	Lr 103

14

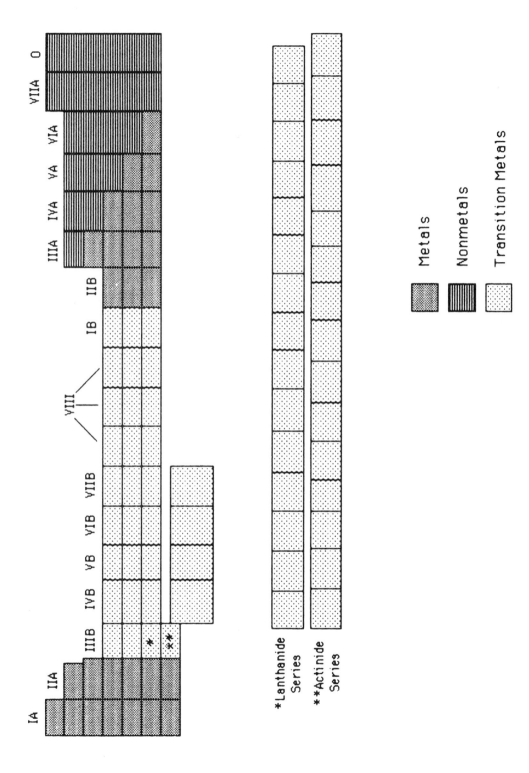

Fig. 2-3. Proportions of metals, nonmetals, and transition metals within the Periodic Table of Elements.

within each group. Serving as a hybrid division between the metals and nonmetals is a group of *metalloid* elements. These elements, such as silicon and germanium, exhibit both metallic and nonmetallic characteristics. Based on these definitions, the Periodic Table can be roughly divided into 77 percent metals, 16 percent nonmetals, and 7 percent metalloids (see Fig. 2-3).

Another peculiar feature of metals centers around the use of valence electrons during bonding. Typically, one to three valence electrons are available for bonding, which eliminates covalent and ionic bonding. Instead, metals pack into tight arrangements of 8 or 12 atoms with overlapping electronic energy levels. There are three standard crystal lattices found in metals:

❖ cubic, body-centered (coordination number 8) as in Fig. 2-4
❖ cubic, face-centered (coordination number 12) as in Fig. 2-5
❖ hexagonal, close-packed (coordination number 12) as in Fig. 2-6

Each of these crystal lattices are formed through a special trait found in metals which is commonly referred to as the *metallic bond*. There are two major theories for explaining the metallic bond: the modified covalent theory and the band theory.

Modified Covalent Theory. The modified covalent theory has its basis on the sharing of electrons between one or more atoms. These

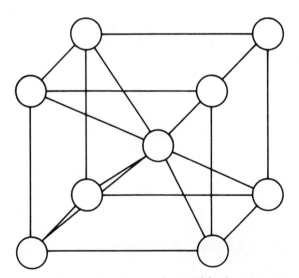

Fig. 2-4. Cubic, body-centered crystal lattice structure.

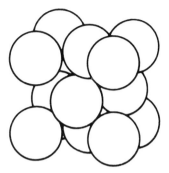

Fig. 2-5. Cubic, face-centered crystal lattice structure.

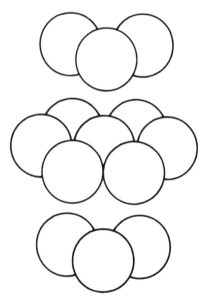

Fig. 2-6. Hexagonal, close-packed crystal lattice structure.

shared electrons then anchor the metal together and generate a metallic bond. This arrangement is similar to a conventional covalent bond and results in the theory's name.

Most of the characteristic properties of metals can be illustrated by the shared electrons of the modified covalent theory. For example, a high electrical conductivity can be demonstrated by the mobile nature of these shared valence electrons.

Band Theory. Certain features of metals (e.g., luster, thermal conductivity, and a more liberal accounting of electrical conductivity) are difficult to explain based solely on the merits of the modified covalent theory. The band theory, however, provides solid explanations for four key stumbling blocks in the modified covalent theory.

+ Decentralized valence electrons. The high coordination numbers found in metallic bonds (8 or 12) use only a small number of valence electrons (one to three) during bonding.

+ Packing atoms in metallic bonds results in overlapping energy levels or bands of electronic energy (hence, the name for this theory).

+ When atomic orbitals combine, they generate molecular orbitals, which in turn are dependent on the number of valence electrons and the number of atomic orbitals.

+ Molecular orbitals increase in number as the number of atoms increases and the energy level difference decreases.

As an illustration of how the band theory satisfies these four metallic bond attributes, consider a typical metal. The average metal consists of an enormous number of atoms with virtually a continuous formation of energy levels. Collectively, these energy levels form an electron band. These bands arise from each atomic orbital and possess a high energy difference. This enormous energy difference precludes the ability for electrons to be promoted from the lower orbital or band to the higher orbitals or bands that are typically half-filled. This inability to transfer electrons to partially filled orbitals results in an energy gap known as a *forbidden zone*.

In addition to partially filled energy bands, the distance between the nuclei also contributes to the bonding process. In the case of internuclear distances, the relative energy of each band is dictated by the proximity of the nuclei. Some metals deal with these variations in distances by, minimizing the forbidden zone through overlapping energy bands.

Band spacing and band filling determines whether a metal is a conductor, nonconductor, or semiconductor. For example,

+ Conductor—partially-filled energy bands.
+ Nonconductor—completely filled (or empty) energy bands with a large forbidden zone.
+ Semiconductor—artificially introduced energy bands through lattice defects.

Due to the partially filled bands found in conductors, introducing an electrical field into a conductor results in the movement of electrons through the lattice in the direction of the introduced field. This effect results in conductivity. This electron movement is found near the top of the filled bands.

METALS

There are two broad categories for metals: *representative* and *transition.* The chief difference between the two categories is the location of the valence electrons. In the representative metals, all of the valence electrons are held in one shell. The transition metals, on the other hand, have their valence electrons scattered throughout more than one shell.

In the Periodic Table, the elements from subgroups IIIB through IB are considered to be transition metals. The representative metals are found in the subgroups IA, IIA, and IIB. Although elements in subgroup IIB are classed as representative metals, they share many of the same characteristics found in the transition metals.

One factor that links the transition metals together is that they are uniform in their physical properties throughout the Periodic Table series. This similarity is attributed to a single electron difference that is found in the second highest shell rather than in the highest valence shell (chromium and copper are exceptions to this rule).

Contrary to the uniform properties of the transition metals, the representative metals demonstrate enormous variety in their physical properties throughout a series. This variability is due to the presence of an electron difference in the outermost valence shell.

Combining either of these metals with another metal or with a minor nonmetal results in the formation of an alloy. In order to be considered an alloy, this metal union must exhibit metallic properties. There are three basic structural groups of alloys:

‡ **Mixture**—insoluble metals form a solid alloy from the combining of each metal's crystals.

‡ **Solution**—atoms from one metal reside inside the crystal lattice of the other component. Replacing some of the solvent atoms with the solute atoms at the lattice points creates a substitutional solid solution. If the atoms reside in holes within the lattice, then an interstitial solid solution is formed.

✛ **Compound**—atoms that comprise the alloy are depicted in atomic ratios. The final formula for these intermetallic compounds is derived from the ratio of the total electron number in the outer shell to the total number of atoms. Generally, this ratio cannot be predicted through conventional valence rules.

Group IA Metals

Lithium, sodium, potassium, rubidium, cesium, and francium are the elements in the group IA metals (see Table 2-1). Also known as the *alkali metals*, the low densities of these group IA metals coined the name *light metals*. Some of the atomic features of the alkali metals include the largest atomic radii, a single electron in the outermost valence shell, and the easy formation of stable positive ions.

Group IIA Metals

Known as the *alkaline earth metals*, the group IIA metals include beryllium, magnesium, calcium, strontium, barium, and radium (see Fig. 2-7). The group IIA metals each have two electrons in their outer shells with a single valence +2. The reactivity of these elements increases with the increase in the element's atomic number.

Group IB Metals

Of all of the metals, this group is the most common. Also known as the *coinage metals*, the group IB metals consists of copper, gold, and silver (see Table 2-2). These transition elements have valence electrons in two shells, arranged with one electron in the outer shell and 18 electrons in the next inner shell. An interesting feature of these group

Table 2-1. Group IA Metals.

Element	Atomic Weight	Atomic Radius, Å
Lithium	6.941	1.52
Sodium	22.9898	1.86
Potassium	39.102	2.31
Rubidium	85.4678	2.44
Cesium	132.9055	2.62
Francium	223	2.70

PERIODS **IIA**

2	Be (4)
3	Mg (12)
4	Ca (20)
5	Sr (38)
6	Ba (56)
7	Ra (88)

Fig. 2-7. The Group IIA metals.

Table 2-2. Group IB Metals.

Element	Atomic Weight	Atomic Radius, Å
Copper	63.546	1.28
Silver	107.868	1.44
Gold	196.9665	1.40

IB metals is the ability of each element to use one or two electrons from the next inner shell for bond formation. This results in $+1$, $+2$, and $+3$ oxidation states in any of the coinage metals.

Group IIB Metals

There are three elements in the group IIB metals: zinc, cadmium, and mercury (see Fig. 2-8). Each member of this representative metals group has two electrons in its outer shell and 18 electrons in the next inner shell. During bonding, coordination numbers of 4 and 6 are found in elemental complexes. The 4-coordinate complexes are tetrahedral, while the 6-coordinate complexes have sp^3d^2 hybridization octahedral structures. Contrary to the alkaline earth metals, these group IIB metals have their reactivity decrease as their atomic numbers increase.

PERIODS **IIB**

4	Zn (30)
5	Cd (48)
6	Hg (80)

Fig. 2-8. The Group IIB metals.

Group III Metals

The group III metals includes subgroups IIIA and IIIB (see Fig. 2-9). Subgroup IIIA consists of boron, aluminum, gallium, indium, and thallium (there is a debate over the inclusion of the nonmetal boron in this group of metals). The subgroup IIIB contains the elements scandium, yttrium, the members of both the lanthanide and the actinide series. All group III elements are trivalent with additional valences demonstrated by members of this group.

The subgroup IIIB metals found within this group are important in superconductivity research due to yttrium and lanthanum (see Fig. 2-10). Even though yttrium isn't a member of the rare earth lanthanide series, its similarity in physical and chemical properties generally includes this element with the rare earths.

Rare Earths. The rare earths are elements with atomic numbers between 57 and 71, inclusive. This subgroup is also known as *lanthanide*

PERIODS **IIIA**

2	B (5)
3	Al (13)
4	Ga (31)
5	In (49)
6	Tl (81)

Fig. 2-9. The Group IIIA metals.

PERIODS **IIIB**

4	Sc (21)
5	Y (39)
6	La (57)*
7	Ac (89)**

*
   Lanthanide Series
**
   Actinide Series

Fig. 2-10. The Group IIIB metals.

series. The major difference between the rare earths and the other metals is that the lanthanide member elements are readily separated by the electrons found in the 4f and 5d subshells. The elements in the lanthanide series become smaller in diameter as their atomic number increases. This size reduction is called the *Lanthanide Contraction.* The relationship that the Lanthanide Contraction has on the rare earth elements leads to many of the members having an atomic size roughly equal to other elements. For example, zirconium and hafnium are the same atomic size (based on hafnium's size reduction from the elements of the lanthanide series). Other shared atomic sizes include niobium and tantalum, molybdenum and tungsten, and yttrium and gadolinium.

Group IV Metals

Germanium, tin, lead, titanium, zirconium, and hafnium constitute the group IV metals (carbon and silicon are nonmetallic members of this same group. See Fig. 2-11. The first three members of this group are found in a subgroup IVA, whereas titanium, zirconium, and hafnium are contained in the subgroup IVB. The metallic properties of the three group IVA metals increase along with their increase in atomic weight. Displaying an oxidation state of +4, the group IVB elements are transition metals with two electrons in their outer shell and ten electrons in the next inner shell.

PERIODS **IVB**

4	Ti (22)
5	Zr (40)
6	Hf (72)

PERIODS **IVA**

4	Ge (32)
5	Sn (50)
6	Pb (82)

Fig. 2-11. The Group IV (IVA and IVB) metals.

Group V Metals

In the group V metals, there are two subgroups: VA and VB. The VA metals are antimony and bismuth, and the group VB elements include vanadium, niobium, and tantalum (see Fig. 2-12). There are three nonmetals in this group: nitrogen, phosphorous, and arsenic, each with a strong attraction between the positive nucleus and the valence electrons. The atoms of these metallic elements have five valence electrons with oxidation states of $+3$ and $+5$.

Group VIB Metals

Chromium, molybdenum, and tungsten are the group VIB metals (see Table 2-3). These transition metals are used exclusively in the production of steel alloys.

Group VIIB Metals

The group VIIB metals are manganese, technetium, and rhenium (see Table 2-4). There are two shells with valence electrons in each of the group VIIB metals. The outer shell contains two of these electrons.

PERIODS **VB**

4	V (23)
5	Nb (41)
6	Ta (73)

PERIODS **VA**

5	Sb (51)
6	Bi (83)

Fig. 2-12. The Group V (VA and VB) metals.

Table 2-3. Group VIB Metals.

Element	Atomic Weight	Atomic Radius, Å
Chromium	51.996	1.25
Molybden-um	95.94	1.36
Tungsten	183.85	1.37

Table 2-4. Group VIIB Metals.

Element	Atomic Weight	Atomic Radius, Å
Manganese	54.9380	1.29
Technetium	98.9062	1.3
Rhenium	186.2	1.37

Group VIII Metals

Contrary to the other metals, the group VIII metals are arranged in three horizontal triads: iron, cobalt, and nickel; ruthenium, rhodium, and palladium; and osmium, iridium, and platinum (see Table 2-5). While the distinguishing features of these three triads is unimportant, one significant attribute of all nine elements is that the ability to lose

Table 2-5. Group VIII Metals.

Element	Atomic Weight	Atomic Radius, Å
Iron	55.847	1.26
Cobalt	58.9332	1.26
Nickel	58.71	1.24
Ruthenium	101.07	1.33
Rhodium	102.9055	1.34
Palladium	106.4	1.38
Osmium	190.2	1.34
Iridum	192.22	1.35
Platinum	195.09	1.38

electrons decreases as the nuclear charge increases. This results in variable oxidation states that change between each of the triads. The last six elements of the group VIII metals—ruthenium, rhodium, palladium, osmium, iridium, and platinum—are also known as the *platinum metals*.

SUPERCONDUCTOR THEORY REQUIREMENTS

In its most basic terms, superconductivity in any metal will be present when a given temperature and magnetic field strength are less than a fixed critical level. An equation for determining these critical levels can be written as:

$$H_c = H_0 [1 - (T/T_c)^2]$$

where,

H_c = critical field strength at T
H_0 = maximum critical field strength at absolute zero
T = temperature
T_c = critical temperature

The results from this equation generate a characteristic curve when the data are plotted. This plotted curve serves as a definition of the barrier between a metal's normal state and its superconducting state. Interpreting the results from this curve can be confusing, however.

For example, assume that a given superconducting metal has an arbitrary critical field of 640 A cm^{-1} at absolute zero. In general terms, this critical field results in a lattice energy of 10^{-7} eV atom^{-1}. Similarly, this same metal has a critical temperature of 7.175 °K at a zero field strength for a lattice energy of 10^{-4} eV atom^{-1}. The relationship between these two interpretations is that there are conflicting forces fighting the superconductivity of a metal. Magnetic field energy is elevating electrons from the superconducting to the normal state, and thermal energy is fighting the superconductive state of the metal. Attempting to explain the superconductivity of any metal must therefore be done in terms of both the metal's critical field strength as well as the metal's critical temperature.

Superconductivity at the Microscopic Level

Every metal exhibits an overall neutral electrical potential lattice structure containing a various number of unbound electrons. Movement of these unbound electrons is governed by the potential of the ionized atoms of the lattice. These unbound electrons occupy a fixed number of energy states based on the energy of the lattice. In a superconductor

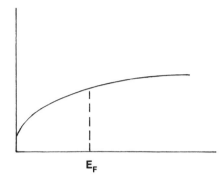

E_F

Fig. 2-13. Normal metal energy state densities for electron energy.

at absolute zero, all of the energy states are occupied below the Fermi energy level. There is no electron energy state occupation above the Fermi energy level. Based on temperature, a Fermi-Dirac energy state occupation distribution can be plotted (see Fig. 2-13).

In a metal's normal state, the Fermi energy level is 10 eV with a temperature of 100,000 °K. This results in the unbound electrons being in a constant state of disarray. Conversely, all of the energy states below the Fermi energy level are occupied at absolute zero. The nature of these two temperature-dependent states, therefore, predominantly effects those electrons near the Fermi energy level.

A Fermi-Dirac energy state occupation distribution shows that energy states that are occupied by unbound electrons change dramatically as the temperature rises above absolute zero (see Fig. 2-14). Furthermore, the number of unbound electrons below the Fermi energy level decreases with this temperature increase.

Another factor that can alter the Fermi-Dirac energy state occupation distribution is the introduction of a voltage pulse. This pulse will give all of the unbound electrons a positive velocity that translates into a current flow. In a superconductor at absolute zero, this current flow will demonstrate a total loss of resistivity.

Based on these alterations in the Fermi-Dirac distribution, the energy spectrum of a superconductor will show a density state gap around the Fermi energy level. This gap ranges in size from 10^{-4} to 10^{-3} eV and is determined by field strength and temperature. An equation for describing this gap is espoused by Bardeen, Cooper, and Shrieffer as:

$$2\epsilon_0 = (3.5kT_c)/2$$

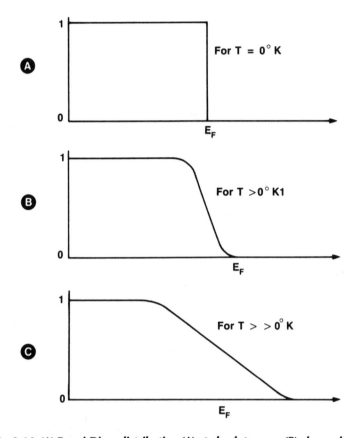

Fig. 2-14. (A) Fermi-Dirac distribution (A) at absolute zero, (B) above absolute zero, and (C) elevated above absolute zero.

where,

ϵ_0 = energy gap
k   = Boltzmann's constant
T_c = critical temperature with a zero field strength

Formation of this energy gap begins with the movement of unbound electrons. These moving electrons generate a positive charge through Coulomb attraction which "bumps" the surrounding lattice ions. In turn, the movement of the lattice ions radiates a disturbance spectrum of acoustic energy called *phonons*. These phonons continue through the lattice via propagation of other electrons. This propagation results in the production of electron attractive forces over a distance factor. The maximum distance for phonon attraction is the *coherence length*.

Electron pair movement effects phonon interaction. This phenomenon is due to the sonic velocities of the electrons nearing a similar speed of the phonons through the lattice that will, in turn, minimize the mutually attractive force holding the electrons in orbit. This is the same effect that was monitored on the surface of magnetic-field superconductors by early experimenters.

When these electrons pair, they form a Cooper pair that has lower energy than the Fermi energy level of unpaired electrons. This reduced energy is determined by the critical field strength that accelerates the pair until phonon interaction is eliminated. All electron pairs are simultaneously effected by the influence of the critical field strength. Therefore, when a given phonon reacts with an electron pair within its coherence length, an increased attraction (greater than the phonon attraction) is imparted. This heightened attraction is the basis for a superconductor's energy gap.

Electron Pair Disruption

There are two factors that can radically alter the essential energy gap of a superconductor:

❖ magnetic field
❖ temperature

The energy gap is steadily reduced as a result of an equally steady increase in a surrounding magnetic field (see Fig. 2-15). Along with this increase in a surrounding magnetic field, the phonon-generated electron pairing is diminished. This effect continues until the energy gap falls to zero.

Temperature can also affect the energy gap. In this case, a rise in temperature will reduce the energy gap (see Fig. 2-16). This reduction in the energy gap is caused by the temperature's influence on the movement of the metal's lattice atoms. Both the amplitude and the frequency of the atom's movement increase which reduces the propagation of phonons. A previous equation can be used for the calculation of this temperature-related energy gap reduction:

$$2\epsilon_0 = 3.5kT_c$$

where,

ϵ_0 = energy gap
k = Boltzmann's constant
T_c = critical temperature

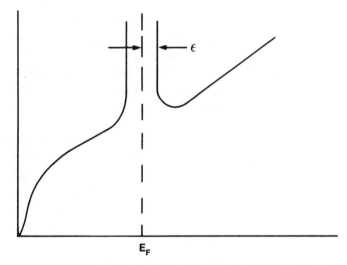

Fig. 2-15. Energy gap formation in a superconducting metal as a function of energy levels.

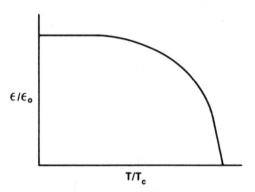

Fig. 2-16. Energy gap curve as a function of critical temperature.

This equation can be modified to show the energy gap's change as related to temperature:

$$\epsilon_0 = 3.2kT_c(1 - T/T_c)^{1/2}$$

where,

ϵ = energy gap
k = Boltzmann's constant
T_c = critical temperature
T = temperature

One final area that influences individual electron pairs is radiation frequencies. Radiations of 10^{11} Hz are roughly equivalent to the energy of phonons that are absorbed by superconductors. Frequencies higher than 10^{11} Hz force the superconductor to its normal state through raising the phonon above the energy gap.

Examining these various means of superconductivity disruption shows that electron pairs can be influenced either together or individually. Joint influences require low energy yields in the guise of magnetic fields and voltages. Discrete influences, however, need high energy yields exceeding the energy gap through temperature and radiation frequency variations.

An interesting footnote to this chapter should now be clearly evident to the reader. By and large, the best electrical conductors result in the formation of poor superconductors. This presumed disparity stems from the lack of a strong acoustic-level lattice-electron interaction in electrical conductors like copper. In fact, it is this absence of interaction which makes copper an ideal electrical conductor. Conversely, excellent high-temperature superconductors lack a measurable degree of electrical conductivity.

IMPORTANT SUPERCONDUCTING METALS

The following pages list 12 superconducting metals and their properties. They are barium, gallium, indium, lanthanum, lead, mercury, niobium, strontium, tin, titanium, vanadium, and yttrium. The figure shows the elemental symbol, the atomic number, the atomic weight, and the electron designations.

BARIUM

Ba 56

137.34

Group IIA Period 6

18, 8, 2

Atomic Number: 56

Atomic Radius, Å: 2.17

Atomic Weight: 137.34

Boiling Point, °C: 1638

Density at 20 °C: 3.5

Ionic Radius, Å: 1.35

Melting Point, °C: 704

GALLIUM

Atomic Number: 49

Atomic Radius, Å: 1.62

Atomic Weight: 114.82

Boiling Point, °C: 2000

Density at 20 °C: 7.36

Ionic Radius, Å: 0.81

Melting Point, °C: 157

INDIUM

In 49

114.82

Group IIIA Period 5

18, 3

Atomic Number: 49

Atomic Radius, Å: 1.62

Atomic Weight: 114.82

Boiling Point, °C: 2000

Density at 20 °C: 7.36

Ionic Radius, Å: 0.81

Melting Point, °C: 157

LANTHANUM

LA 57

138.9055

Group IIIB Period 6

18, 9, 2

Atomic Number: 57

Atomic Radius, Å: 1.88

Atomic Weight: 138.9055

Boiling Point, °C: 880

Density at 20 °C: 6.15

Ionic Radius, Å: 1.15

LEAD

Pb 82

207.2

Group IVA Period 6

32, 18, 4

Atomic Number: 82

Atomic Radius, Å: 1.75

Atomic Weight: 207.2

Boiling Point, °C: 1750

Density at 20 °C: 11.34

Ionic Radius, Å: 0.84

Melting Point, °C: 327.4

MERCURY

80

Hg

200.59

Group IIIB Period 6

32, 18, 2

Atomic Number: 80

Atomic Radius, Å: 1.55

Atomic Weight: 200.59

Boiling Point, °C: 356.57

Density at 20 °C: 13.546

Ionic Radius, Å: 1.1

Melting Point, °C: −38.87

NIOBIUM

Nb

41

92.9064

Group VB Period 5

12, 1

Atomic Number: 41

Atomic Radius, Å: 1.43

Atomic Weight: 92.906

Boiling Point, °C: 4930

Density at 20 °C: 8.57

Ionic Radius, Å: 0.69

Melting Point, °C: 2487

STRONTIUM

Sr

38

87.62

Group IIA Period 5

8, 2

Atomic Number: 38

Atomic Radius, Å: 2.15

Atomic Weight: 87.62

Boiling Point, °C: 1384

Density at 20 °C: 2.60

Ionic Radius, Å: 1.13

Melting Point, °C: 770

TIN

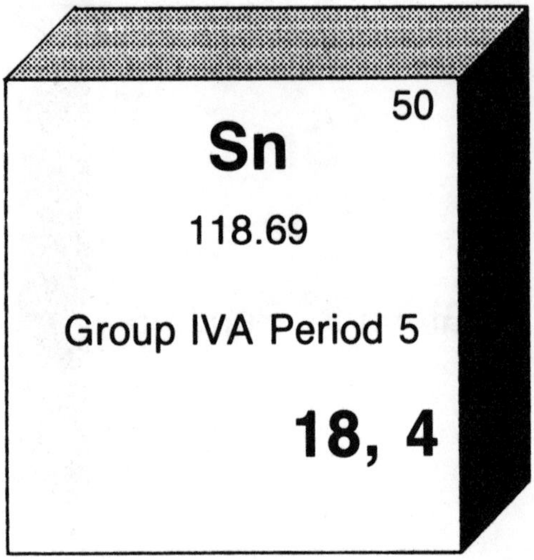

Sn
50
118.69
Group IVA Period 5
18, 4

Atomic Number: 50

Atomic Radius, Å: 1.4

Atomic Weight: 118.69

Boiling Point, °C: 2337

Density at 20 °C: 7.31

Ionic Radius, Å: 0.71

Melting Point, °C: 231.9

TITANIUM

Ti 22

47.90

Group IVB Period 4

10, 2

Atomic Number: 22

Atomic Radius, Å: 1.46

Atomic Weight: 47.90

Boiling Point, °C: 3260

Density at 20 °C: 4.49

Ionic Radius, Å: 0.64

Melting Point, °C: 1812

VANADIUM

V   23

50.9414

Group VB Period 4

11, 2

Atomic Number: 23

Atomic Radius, Å: 1.31

Atomic Weight: 50.942

Boiling Point, °C: 3375

Density at 20 °C: 5.96

Ionic Radius, Å: 0.40

Melting Point, °C: 1730

YTTRIUM

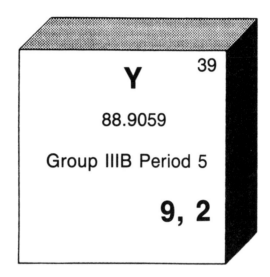

Atomic Number: 39

Atomic Radius, Å: 1.80

Atomic Weight: 88.9059

Boiling Point, °C: 1500

Density at 20 °C: 4.47

Ionic Radius, Å: 0.93

3

Superconductivity Microscopic Theory

Trying to understand superconductivity on the basis of a strict interpretation of contemporary inorganic valence electron behavior is difficult, at best. One simplified attempt at addressing this conflict would be to state that ordinary valence electrons take on a "superelectron" configuration in superconductors. However, this explanation lacks the sophistication that is necessary for answering all of the properties exhibited by the superconducting state. In order to account for superconductivity in metals more thoroughly, a solid knowledge of quantum mechanics is required.

SIX SUPERCONDUCTOR PROPERTIES

There are six generally endorsed properties of superconductivity that must be satisfied by any proposed theory:

1. **Lack of resistivity**. Superconductors display zero resistance to direct currents and low-frequency alternating currents. This is in direct conflict with the superconductor's resistivity to optical frequencies, however. Based on Maxwell's equations, the resistivity of a superconductor is identical in both its normal state and its superconducting state. A typical superconductor

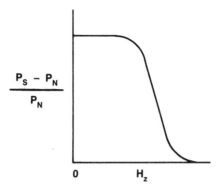

Fig. 3-1. Typical metal infrared reflection coefficient curve.

loses its resistivity as frequencies enter the microwave regions of the electromagnetic spectrum (see Fig. 3-1).

2. **Meissner effect**. Linked to a superconductor's lack of resistivity, the formation of a magnetic field, or the *Meissner effect*, is a fundamental property of superconductivity. (This concept lacks an adequate interpretation at the elementary level.)

3. **Crystal lattice structure**. Both the lattice symmetry and the lattice spacing remain constant throughout a superconductor's transformation from its normal state into its superconducting state. Even properties that are derived from the crystal lattice structure, such as specific heat and the Debye temperature, remain constant through all changes in state.

4. **Electron distance coherence**. Boundaries of change between the normal and superconducting states is a gradual process without a sharp definition. This effect leads to the assumption that superconductors have electrons that exhibit a long-range order. In other words, the presence of superconducting electrons gradually dissipates beyond a fixed superconducting/normal state boundary based on the coherence distance for the phonon-activated electrons.

5. **Specific heat**. Cooling a superconductor below its critical temperature without the influence of a magnetic field increases the specific heat of the metal without an appreciable formation of latent heat. If this same superconductor is forced back into a normal state through the introduction of a magnetic field, the

lattic specific heat doesn't change, but the conduction electrons are altered.

6. **Electron isotope change**. Two independent research groups (Maxwell and Reynolds, and Serin, Wright, and Nesbitt) measured different critical temperatures for various isotopes of the same element. Generally, it was found that the critical temperature for these isotopes is inversely proportional to the square root of the isotope's mass. One conclusion from this observation is that the crystal lattice structure determines the properties of the electrons found in these elemental isotopes.

One lesser quality of superconductors that is readily observed is the occurrence of a radiation absorption curve. This same event is seen in semiconductors. In the case of semiconductors, this curve is caused by an energy gap that forms between the full (outer) energy band (this band is also known as the *valence band)* and the empty (next inner level) energy band (or *conduction band*). High frequency optical radiation directed on this energy gap forces an electron to move from the valence band to the conduction band. Furthermore, the photons used for this excitation are absorbed by the semiconductor.

A similar explanation can be hypothesized for the radiation absorption curve found in superconductors. By using the semiconductor model, the optical radiation must be absorbed to a larger degree at the moment the photons excite the electrons through the energy gap. Based on these assumptions, the energy gap found in a typical superconductor can be expressed as:

$$10^{-4} \text{ eV}$$

where optical frequencies greater than 10^{11} Hz have been applied to the superconductor.

Further credence can be lent to support the hypothesis of an energy gap in superconductors by examining the specific heat of certain metals as they enter and exit the superconducting state. Much of these data are obtained from the conduction electrons in the superconducting state. As verification of the existence of an energy gap, the conduction electrons should be proportional to the following equation in a low-temperature superconductor:

$$e^{-b/kT}$$

where,

$$k = \text{Boltzmann's constant}$$

Electrons are excited across the energy gap with an increase in the temperature. This results in the absorption of an energy level equal to the energy gap:

$$E_g$$

Applying these two relationships to a given temperature (T) results in the electrons in the conduction band to be proportional to:

$$e^{-E_g/2kT}$$

Additionally, the absorption of thermal energy is proportional to:

$$E_g e^{-E_g/2kT}$$

Therefore, the specific heat is proportional to the derivative of the energy level as a factor of temperature. This results in the variation of specific heat being virtually exponential:

$$(1/T^2)e^{-E_g/2kT}$$

PHONON PRODUCTION

An area of conventional metal interaction that lacks an interpretation within this definition of a superconductor's energy gap is *electrostatic coulomb repulsion*. In conventional metals, coulomb repulsion is a natural event between electrons. In a superconductive metal, however, coulomb repulsion fails to suggest the development of an energy gap. As has been previously stated, a superconductor's energy gap must be derived from a minimal interaction at the atomic level. The best model for examining this energy gap formation, therefore, is an electron-crystal lattice structure interaction.

An electron-crystal lattice structure interaction is initiated by a thermally induced lattice vibration. This vibration forms a wave action through the crystal lattice structure and causes the electrons to be scattered. These scattered electrons perpetuate the vibration through the excitation of a phonon.

Just as lattice-vibrated electrons are able to generate phonons, neighbor electrons are able to absorb phonons. One worker studying this electron-phonon-electron interaction was Fröhlich in 1950. Frohlich theorized that this electron-phonon-electron interaction actually linked the two electrons together through the phonon. In this application, the phonon served as an energy transmitter between the two electrons. Fröhlich's scenario depicted the lattice vibrations as scattering electrons which, in turn, formed phonons. These phonons were then absorbed by other electrons (see Fig. 3-2). This phonon excitation-absorption cycle, Fröhlich stated, would generate an energy gap bordering on the speculated 10^{-4} eV level.

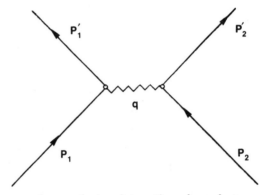

Fig. 3-2. Electron-phonon-electron interaction where electron p1 has emitted the phonon q causing an interaction with electron p2.

Each of Fröhlich's assertions can be studied mathematically as follows:

Momentum of a phonon:

$$q = hv/s$$

where,

q = momentum of a phonon
h = Planck's constant
v = frequency of a phonon
s = velocity of sound

Momentum of excitation electron:

$$p_1 = p_1' + q$$

where,

p_1 = momentum prior to electron scattering
p_1' = momentum following scattering
q = momentum of the emitted phonon

Momentum of absorption electron:

$$p_2 = p_2' + q$$

where,

p_2 = momentum following absorption
p_2' = momentum prior to absorption
q = momentum of the emitted phonon

Solving for q:

$$q = p_1 - p_1'$$
$$q = p_2 - p_2'$$
$$p_1 + p_2 = p_1' + p_2'$$

This series of equations proves Froħlich's assumption that the momentum of a vibrated electron is conserved. Based on this theoretical proof, however, it would be incorrect to presume that the energy levels in the intermediate state (the state between phonon emission and absorption) are conserved. This action, in which the energy level is not conserved due to a time/energy uncertainty equation, is called a *virtual action*. Likewise, a virtual action cannot happen unless there is another electron in close proximity to the emitting electron that is capable of absorbing the phonon.

Typically, the difference between the energy levels of the emitting electron before and after the virtual release of the phonon is less than hv. In other words, there is an attraction between the two electrons. It is this attractive interaction that accounts for superconductivity in metals. Furthermore, this electron-phonon-electron interaction also aids in illustrating the cause for a superconductor's poor normal-state conductivity. This can be interpreted to mean that a *strong* electron-phonon-electron interaction represents a good *superconductor*, and a *weak* interaction represents a good *conductor*.

COOPER'S ELECTRON THEORY

Building on the electron-phonon-electron interaction, L. N. Cooper espoused an explanation for the addition of two electrons to a superconductor. Essentially, Cooper stated that the electrons will bond with a resultant energy less than two times their Fermi energy $(2\epsilon_F)$, provided the electrons' momenta is greater than the radius of an occupation sphere (p_F).

$$p_F = \sqrt{2m \; \epsilon_F}$$

where,

p_F = radius of momentum space sphere

$m\epsilon_F$ = Fermi energy rest mass

Cooper used quantum mechanics as proof of his assertions. Beginning with the probability that two given electrons with given momenta would be at a given set of coordinates:

p_1 = momentum of first electron

p_2 = momentum of second electron

(x_1, y_1, z_1) = coordinates of first electron

(x_2, y_2, z_2) = coordinates of second electron

$\psi(x_1, y_1, z_1, p_1)$ = first electron wave function

$\psi(x_2, y_2, z_2, p_2)$ = second electron wave function

$\phi(x_1, y_1, z_1, p_1, x_2, y_2, z_2, p_2)$ = two-electron wave function

$\phi(x_1, y_1, z_1, p_1, x_2, y_2, z_2, p_2)$ = $\psi(x_1, y_1, z_1, p_1)\psi(x_2, y_2, z_2, p_2)$

or,

$$\phi(p_1, p_2) = \psi(p_1)\psi(p_2)$$

In these equations, the first and second electron wave functions (ψ) are Bloch plane wave functions lacking any interaction. By dispersing the electrons through changes in momenta the wave functions will be altered and a new wave function will be formed:

$$\Phi \; (x_1, y_1, z_1, x_2, y_2, z_2) = \overset{1,2}{\Sigma} \; a_{1 \; 2} \; \psi(P_1) \; \psi(p_2)$$

Based on this new wave function the continually dispersing electrons can be located with the probability $/a_{1 \; 2}/^2$. A more mathematically sound means of looking at these two electrons is through the use of any electron pair i and j. This conversion results in a probability of $/a_{ij}/^2$ and a wave function:

$$\Phi \; (x_1, y_1, z_1, x_2, y_2, z_2) = \underset{i, j}{\Sigma} \; a_{i \; j} \; \psi(p_1) \; \psi(p_2)$$

The dispersing of the two electrons does not alter the conservation of momentum. Therefore,

$$p_i + p_j = P$$

A negative potential energy is formed from the attractive interaction between the two electrons. Over a fixed time interval, each electron's energy decreases proportionally with the number of dispersions due to a negative potential energy time-average.

While the interaction between these two electrons is attractive, a more detailed definition was forwarded by Cooper. Working with the energy levels for the initial and intermediate states found in a virtual phonon emission and absorption, the momenta for electron pairs that satisfies $p_i + p_j = P$ are found within two small bands. The number of these electron pairs is proportional to the volume of the P space of the electron ring. When $P = 0$, this ring forms a sphere of thickness Δ_p. Therefore, only by pairing electrons with equal and opposite momenta will the maximum number of dispersions with the maximum decrease in energy be initiated (see Fig. 3-3).

A supplemental notation to this conclusion states that the greatest decrease in energy is possible when the two electrons have opposite spin. This point is verified in the ground state of a hydrogen molecule. These variations in spin are indicated:

p^l = an up-spin electron with momentum p

$-p^l$ = a down-spin electron with momentum $-p$

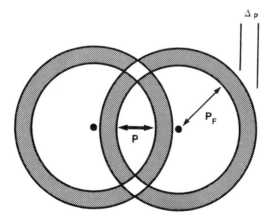

Fig. 3-3. Cooper pair formation, where pair production is proportional to the shell intersection areas.

Applying this supplemental spin reference to the Φ wave function:

$$\phi(p_i\uparrow, -p_i\downarrow) = \psi(p_i\uparrow]\,\psi(-p_i\downarrow)$$
$$\Phi\,(x_1, y_1, z_1, x_2, y_2, z_2) = \sum_i a_i\,\phi(p_i\uparrow, -p_i\downarrow)$$

Note: The dual electron designation i and j has been ammended to i based on the variations in electron spin.

Cooper termed this wave-function description a *Cooper pair*. Additionally Cooper found that the formation of an electron pair featuring equal and opposite momenta results in a decrease in the potential energy, which is higher than the kinetic energy's increase above $2\epsilon_F$. The following equations illustrate this point:

$$W = 2\sum_i /a_i/^2\,(p_i^2/2m)$$
$$p > p_F$$
$$\sum_i /a_i/^2 = 1 \;(\textit{Note: For a normalized } \Phi \text{ wave function.})$$
$$2p_F^2/2m = 2\epsilon_F$$

Due to this relationship between a Cooper pair and total energy, Cooper arrived at the following conclusion: If two electrons enter a Φ wave function state with a range $\Delta_p = mhv_l/\epsilon_F$, then the total energy is less than the total energy of two electrons entering states higher than p_F and lacking any interaction.

THE BARDEEN, COOPER, AND SCHRIEFFER THEORY ———————

A significant advance over Cooper's theory was the application of this modest two-electron conclusion to all of the electrons found within a superconductor. In 1957, three scientists—J. Bardeen, L. N. Cooper, and J. R. Schrieffer—demonstrated a theory that examined the total atomic structure of a superconductor.

The Bardeen, Cooper, and Schrieffer theory begins with two general assumptions:

✤ Only Cooper pair interactions are significant in a superconductor.
✤ All non-Cooper pair electrons in a superconductor only serve as limiters on the states into which the interacting pair can be dispersed. *Note*: Pre-interaction state occupation and the Pauli principle serve as the limiting agents in this assumption.

The first step that the Bardeen, Cooper, and Schrieffer (BCS) theory addresses is the possibility of adapting Cooper's calculations for a single Cooper pair to multiple Cooper pairs within the same metal. This conversion is readily accomplished through a minor reworking of the Φ wave function equation:

$$\Phi\ (x_1,\ y_1,\ z_1,\ x_2,\ y_2,\ z_2)\ \Phi\ (x_3,\ y_3,\ z_3,\ x_4,\ y_4,z_4)$$
$$\ldots\ \Phi\ (x_{n_s-1},\ y_{n_s-1},\ z_{n_s-1},\ x_{n_s},\ y_{n_s},\ z_{n_s})$$

In order to simplify this equation, a new variable, r, can be substituted for the position coordinates.

$$\Phi\ (r_1,\ r_2)\ \Phi\ (r_3,\ r_4)\ \ldots\ \Phi(r_{n_s-1},\ r_{n_s})$$

A new wave function, Ψ_G, is used for representing these multiple electron pairs.

$$\Psi_G\ (r_1,\ r_2,\ \ldots\ r_{n_s})\ =\ \Phi(r_1,\ r_2)\ \ldots\ \Phi(r_{n_s-1},\ r_{n_s})$$

As shown by the Ψ_G wave function, the total number of electron pairs can be expressed by $n_s/2$. With the groundwork for multiple electrons established by the Ψ_G wave function, determining the probability of locating one electron at r_{n_s-1}, while a second electron is at r_{n_s} is possible, exclusive of their momenta. There is a limit, however, on the number of electrons that can form Cooper pairs.

By reviewing the previously explained spin momenta equations, an explanation for this electron limit can be derived. A typical rendering of electron dispersion between $p_i\uparrow,\ -\ p_i\downarrow$ and $p_j\uparrow,\ -\ p_j\downarrow$ requires that $p_i\uparrow,\ -\ p_i\downarrow$ must be occupied and $p_j\uparrow,\ -\ p_j\downarrow$ be empty. As electrons form Cooper pairs, the likelihood that $p_j\uparrow,\ -\ p_j\downarrow$ will remain empty becomes increasingly small. Thus the dispersion process becomes limited and the negative potential energy decreases. Finally, a level is reached where the potential energy can't offset the kinetic energy and the total energy will not be lower by forming Cooper pairs. Determining the configuration that provides the lowest energy or ground state is set by the probability h_i.

$$\Delta\ =\ 2hv_L\ \exp[-\{\ N\ (\epsilon_F)V\}^{-1}]$$
$$h_i\ =\ \tfrac{1}{2}[1\ -\ (\epsilon_i\ -\ \epsilon_F)/\{(\epsilon_i\ -\ \epsilon_F)^2\ +\ \Delta^2\}^{\frac{1}{2}}]$$

where,

$$h_i = \text{probability}$$
$$\epsilon_i = p_i^2/2m$$
$$v_L = \text{phonon frequency}$$
$$V = \text{dispersion interaction}$$
$$N(\epsilon_F) = \text{state density at the Fermi energy}$$

A ground state occurs when all electrons with a given momenta are coupled in Cooper pairs with opposite momentum and spin. This form of a ground state is called a *condensed state*—where the electrons form a lower energy state similar to that found when a gas condenses into a liquid.

Even though the momentum of the electrons is constantly changing, the total energy of the interacting electron pairs remains steady. In a time-dependent function, the two-electron wave functions all exhibit the same frequency of oscillation, resulting in a phase relationship. An application of this expression states that wave function Φ is a coherent mixture of the wave function Ψ_G (Rose-Innes & Rhoderick, 1978).

Applying the Bardeen, Cooper, and Schrieffer Theory

There are three areas for application of the Bardeen, Cooper, and Schrieffer theory:

✛ the energy gap
✛ critical temperature
✛ critical magnetic field

By applying energy to a Cooper pair, the two electrons can be broken apart into two separate units lacking opposite and equal momenta. These "free" electrons are referred to as quasi-particles. Each of these quasi-particles has a well-defined momenta. The Bardeen, Cooper, and Schrieffer theory states that the amount of energy required to break a Cooper pair into these quasi-particles is:

$$E = E_i + E_j$$

or,

$$E_i = \{(\epsilon_i - \epsilon_F)^2 + \Delta^2\}^{1/2}$$
$$E_j = \{(\epsilon_j - \epsilon_F)^2 + \Delta^2\}^{1/2}$$
$$E = \{(\epsilon_i - \epsilon_F)^2 + \Delta^2\}^{1/2} + \{(\epsilon_j - \epsilon_F)^2 + \Delta^2\}^{1/2}$$

From this equation, the minimal amount of energy needed to form two quasi-particles is 2Δ for $\epsilon_i = \epsilon_j = \epsilon_F$. Therefore, the energy gap of a superconductor is 2Δ with frequency v absorbed when $hv > 2\Delta$.

In addition to this means of forming an energy gap, the Bardeen, Cooper, and Schrieffer theory also relates a secondary formation method. If pair states are not available to the remaining Cooper pairs, then the number of dispersions will be reduced with a similar decrease in electron binding energy. Therefore, the total energy increases and an energy gap will be generated. In both cases, the loss of the electron binding energy is linked with the formation of an energy gap (see Fig. 3-4).

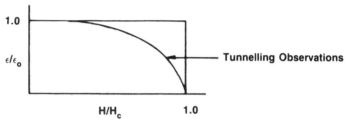

Fig. 3-4. Energy gap difference as a function of magnetic field strength. Actual tunnelling observations are used for the formation of the illustrated curve. A theoretical curve shows a sharp decline in the energy gap for the magnetic field.

Another means of breaking Cooper pairs into smaller quasi-particles is via thermal vibration due to an increase in temperature. During this rise in temperature, the energy gap decreases. This energy gap decrease phenomenon is similar to the last-pair states, as mentioned above, that are available to Cooper pairs. Consequently, the reduction in electron pair interaction correlates with a reduction in the energy gap. Therefore, as the temperature increases, the number of quasi-particles increases and the energy gap diminishes. Finally, a temperature is reached where the energy gap is reduced to zero. This temperature, T_c, represents the critical temperature for the superconductor. A relationship between the energy gap and the critical temperature at absolute zero can be written:

$$E_g(O) = 2\Delta(O)$$

or,

$$2\Delta(O) = 3.5k\, T_c$$

By replacing Δ, the critical temperature can be expressed as:

$$3.5kT_c = 4hv_L \exp[-\{N(\epsilon_F)V\}^{-1}]$$

Specific heat rises abruptly as temperature nears the critical temperature. At temperatures above the critical temperature, however, there is no electron-generated specific heat, which results in a sharp decline in the specific heat as the temperature continues to increase. Likewise, based on the decline in the energy gap, the total energy is equal whether the temperature is rising above or below the critical temperature. This equivalency results in an absence of latent heat.

In terms of the critical magnetic field of a superconductor, the Gibbs free energy densities for the metal in both its normal and superconducting states at absolute zero are used:

$$\tfrac{1}{2}\,\mu_O H_c^2 = g_n - g_s$$

These values represent the binding energy of a normal-state metal and a superconducting-state metal where Cooper pairs are present. Due to this representation, the above equation can be adapted with previously known values:

$$\tfrac{1}{2}\,\mu_O H_O^2 = (g_n - g_s)_O$$
$$\tfrac{1}{2}\,\mu_O H_O^2 = \tfrac{1}{2}\,N(\epsilon_F)[\Delta(0)]^2$$

Finally, the relationship of $\Delta(0)$ to the critical temperature can be added for forming a "law" of corresponding states:

$$H_O^2/T_c^2 = \frac{0.47\,[\tfrac{2}{3}\,\pi^2\,N(\epsilon_F)k^2]}{\mu_O}$$

RESISTIVITY

The most exciting attribute of the superconductor is its ability to conduct current with zero resistance. At the microscopic theory level, the reason for this loss of resistivity can be easily explained through collective Cooper-pair momentum.

Prior to this point, the discussion of Cooper pairs failed to consider the overall momentum of *all* of these electron pairs. This collective momentum can be easily added to the previous equations dealing with pair momentum:

$$\left[\left(p_i + \frac{p}{2}\right)\uparrow, \left(-p_i + \frac{p}{2}\right)\downarrow\right]$$

where,

$$p = \text{total pair momentum}$$

Like the electron momentum equations, this collective pair momentum equation follows a dispersion from i to j. Furthermore, this *collective* momentum is equivalent to the previously considered zero momentum. Therefore, the Φ wave function can be modified to accommodate this new collective momentum interpretation (this exercise is left to the reader).

Essentially, a current in a superconductor is then carried by the electron pairs with a collective momentum of P. This is in direct contrast with current conduction in a normal-state conducting metal. In this case, the current encounters resistance from the electrons (or holes) randomly dispersing towards the electric field. Contrary to this conventional conducting dispersion, Cooper pairs in a superconductor disperse at a fixed momentum P. Likewise, the current is conducted without variation. This conduction process is termed *zero resistivity*.

This zero resistivity can be disrupted, however. As previously illustrated, once the Cooper pairs begin to break into quasi-particles, the collective momentum in the superconductor decreases. Only a fixed energy level (2Δ) is able to instigate this quasi-particle formation and its subsequent loss of zero resistivity. Typically, low current level or densities are unable to impart this energy level to the Cooper pairs, and zero resistivity in the superconductor prevails. There are current densities, however, that can cause the breakdown of the Cooper pair.

The current density at which resistance forms in a superconductor is called the *critical current density*.

$$j_c = \frac{en_s\Delta}{p_F}$$

In this expression, the total momentum is proportional to the current density. In other words, there is a current density where the dispersion is coupled with a change in momentum. This change in momentum forces the breakdown of the Cooper pair into quasi-particles. In turn, these quasi-particles increase the formation of resistance. Therefore, an increase in the current density above the critical current density causes the superconductor to lose its zero resistivity as the Cooper pairs are replaced by quasi-particles.

This elaborate series of theoretical interpretations conducted at the microscopic level by Bardeen, Cooper, and Schrieffer satisfied all of the questions raised by the early studies of superconductor thermodynamics.

✜ Why are superconductor transition temperatures so low?

✜ How do electrons form ordered pairs?

✜ Why doesn't Coulomb repulsion effect electrons in a superconductor?

✜ How are currents carried through a superconductor without resistance?

From this fundamental understanding of the quantum mechanics associated with superconductors, the actual thermodynamics of superconductors is more easily appreciated. In turn, this comprehension of the thermodynamics of superconductors will lead to a successful implementation of superconductivity in practical applications.

4

Superconductor Thermodynamics

When a metal makes its known change in state from a normal conductor into a superconductor, there is a sudden thermodynamic transition. In a controlled environment, this second-order transition is devoid of latent heat. There is a telltale signature, however, in the "movement" of the specific heat at the thermodynamic transition point.

As a metal is cooled from its normal state into a superconducting one, the specific heat takes a dramatic vertical rise at the transition point between these two states (see Fig. 4-1). At temperatures below this transition temperature, the specific heat begins to gradually drop below the anticipated normal specific heat curve. Even though this characteristic climb in the specific heat is similar for all superconductors, lattice impurities in the metal will affect the slope of the curve and the definition of the transition point.

A useful extension from this rise in the specific heat of a metal entering a superconducting state is the interpretation of the entropy of the superconductor. Beginning with an expression of the entropy of a superconductor at absolute zero,

$$S = \int_{0}^{T} C_{v}(T')/T' dT'$$

at any given temperature below the transition point, the integration of

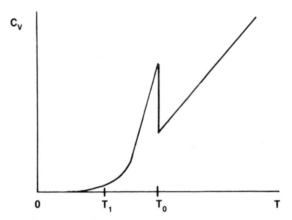

Fig. 4-1. Specific heat for a superconductor makes a rapid jump at the state's transition temperature. Below this transition temperature, the specific heat is reduced to zero.

entropy for a superconductor will be significantly lower than for the same metal in a normal state.

Therefore, based solely on a thermodynamic examination of a given superconductor, two conclusions can be drawn:

✤ The superconducting state is highly ordered. This feature relates to the lack of entropy at the low temperatures of superconductivity.
✤ The specific heat of the superconductor is important in its indirect contribution to the entropy of a metal.

Obviously, these two conclusions have been previously identified through the microscopic theory (see Chapter 3). The fact that these same conclusions were reached through a different, independent means, however is significant in solidifying a firm understanding in superconductivity. In light of this supplemental support being derived through applications of thermodynamics, a brief review of the three laws of thermodynamics is appropriate.

THE LAWS OF THERMODYNAMICS

Thermodynamics is the study of energy transformations that occur within a system as well as the energy transfers that occur between a system and its surroundings. When dealing with the thermodynamics of any system, there are two important points to remember:

✤ Systems move toward a state of minimum potential energy.
✤ Systems move toward a state of maximum disorder.

Basically, thermodynamics can be thought of as the amount of work, in terms of energy, that can be acquired from a system when a portion of the energy is transformed into heat and transferred to the surroundings. This essential element of thermodynamics is controlled by the feature that the maximum amount of energy available for work is limited by the amount of energy released by the system. More specifically, the total released energy cannot be completely utilized as total work energy.

First Law of Thermodynamics

The *First Law of Thermodynamics* or *Law of Conservation of Energy* states:

The total amount of energy in the universe is constant.

In order to prove the First Law, a system with an internal energy state is added to an amount of heat energy, which in turn performs work energy to the surroundings. This forces the system to acquire a new internal energy state. This same statement can be illustrated with the equation:

$$E_2 = E_1 + (q - w)$$

where,

E_2 = new system internal energy state
E_1 = initial system internal energy state
q = heat energy
w = work energy

The initial and new internal energy states for the systems can be expressed as:

$$E_2 - E_1 = q - w$$

In turn, the relationship $E_2 - E_1$ represents the change in energy for the system or Δ_{sys}.

$$\Delta_{sys} = E_2 - E_1$$

This relationship means that the change in energy state for the system is equal to the difference between the heat energy transferred to the system and the work energy transferred from the system. In both of these energy transfer cases, the energy is moving between the system and its surroundings.

$$\Delta_{sys} + \Delta_{sur} = \phi$$

where,

$$\Delta_{sur} = \text{change in energy for the surroundings}$$

Therefore, as illustrated in this final equation, the total amount of energy in the universe is constant.

Before the Second Law of Thermodynamics is discussed, there are three important supplements that must be made to the First Law: enthalpy change, free energy change, and entropy change.

Enthalpy Change. Pressure/volume (or PV) work forms a significant influence over the energy of a system. This influence is usually represented through the addition of the term $P\Delta V$ to a total internal energy change ΔE. This addition generates an expression for the total amount of heat energy that can be transferred to the surroundings. The term ΔH is used for this representation of the change in heat content. Another name for this change in heat content is the change in *enthalpy*. By combining all of these terms:

$$\Delta H = P\Delta V + \Delta E$$

At a constant pressure:

$$w = P\Delta V$$

Applying this value to the First Law:

$$\Delta E = q - P\Delta V$$

Finally, based on enthalpy change:

$$\Delta H = q$$

Therefore, at a constant pressure, the *enthalpy change* can be stated as *the quantity of heat absorbed by a reaction at a constant pressure*. An extension of this statement can be made for constant volumes as:

$$\Delta H = \Delta E$$

Depending on whether the enthalpy change is positive or negative, the heat energy can be either transferred or absorbed by the system from its surroundings. In these cases, a negative enthalpy change results in an exothermic reaction where the heat energy is transferred to the surroundings. On the other hand, a positive enthalpy change results in an endothermic reaction where the heat energy is absorbed by the system.

Measuring the enthalpy change for pure substances at a temperature of 298.15 $^\circ$K at one atmosphere is called a *standard state condition*. The symbol used for indicating standard state condition measurements is ΔH°_{298}. Furthermore, if one mole of the test substance is used for the standard state condition determination, the enthalpy change is called a *standard molar enthalpy of formation*. The symbol used for standard molar enthalpy of formations is ΔH°_{f298}. One point to remember regarding enthalpy change is that the standard molar enthalpy of formation for a free element is zero in its most stable form.

Free Energy Change. A reaction at a constant temperature and pressure will yield a maximum amount of useful work known as the *Gibbs free energy change* or ΔG. A negative ΔG represents a reaction that occurs spontaneously at a constant temperature and a constant pressure. Conversely, a positive ΔG indicates a reaction where the reverse reaction is spontaneous at a constant temperature and a constant pressure. Therefore, the *Gibbs free energy change* states *any system that is at equilibrium has a ΔG equal to zero.*

Because G is a state function, calculations for the standard state of free energy change, ΔG°, can be derived from the standard molar free energies of formation (ΔG°_f), similar to the method used for ΔH° values.

Entropy Change. The amount of disorder within a system is measured as the entropy, S, of the system. The entropy change for a system is represented as ΔS. The ΔS is equal to the difference between the entropies of the products for a given reaction and the entropies of the reactants. Therefore, if ΔS is positive, the system is increasing in disorder. Similarly, if ΔS is negative, then the system is moving towards an ordered state. This involvement between entropy and system disorder indicates that entropy change measures a system's inclination towards increasing its disorder.

S is a state function as are all of the factors previously considered (E, H, and G). This relationship permits the incorporation of *entropy change* with enthalpy change which states that *reactions move towards a state of mimimum energy and maximum disorder.* Another way of stating this relationship is that reactions move towards a state of $-\Delta H$ and $+\Delta S$.

At a given temperature Kelvin, changes in free energy, enthalpy, and entropy can be described as:

$$\Delta G = \Delta H - T\Delta S$$

This equation is also known as the Gibbs-Helmholtz equation. By using the Gibbs-Helmholtz equation, the spontaneity for any reaction can be determined. In this application, $T\Delta S$ acts as a correction term between ΔG and ΔH. Examples of some reactions that can be evaluated with the Gibbs-Helmholtz equation include:

Exothermic Reactions: Spontaneity $= -\Delta H + T\Delta S = -\Delta G$
Not Spontaneous $= \Delta H +)-T\Delta S)$

Endothermic Reactions: Spontaneity $= \Delta H + (-T\Delta S) = \Delta G$
Not Spontaneous $= \Delta H + T\Delta S$

Note: The value T is always positive, so ΔS establishes the sign.

Second Law of Thermodynamics

The *Second Law of Thermodynamics* states:

During a spontaneous reaction, the entropy of the universe increases.

The Second Law is best used through the Gibbs-Helmholtz equation:

$$\Delta G = \Delta H - T\Delta S$$

Solving this equation for a system:

$$\frac{\Delta G_{sys}}{T} = \frac{\Delta H_{sys}}{T} - \Delta S_{sys}$$

By substituting a variable for the entropy change of the surroundings in relation to the enthalpy change for the system at a constant pressure and a constant temperature:

$$\Delta S_{sur} = -\frac{\Delta H_{sys}}{T}$$

The Gibbs-Helmholtz equation for a system becomes:

$$\frac{\Delta G_{sys}}{T} = -\Delta S_{sur} - \Delta S_{sys}$$

A change in the sign:

$$\frac{-\Delta G_{sys}}{T} = \Delta S_{sur} + \Delta S_{sys}$$

Based on the equation:

$$\Delta S_{universe} = \Delta S_{sur} + \Delta S_{sys}$$

This relationship exists:

$$\Delta S_{universe} = \frac{-\Delta G_{sys}}{T}$$

In this case, a positive change in entropy (increased disorder) in the universe results in a negative change in the free energy of the system (a spontaneous reaction).

Therefore, as shown with this final equation, any spontaneous change results in an increase in the change of entropy of the universe. In other words, the system is capable of executing useful work on its surroundings.

Third Law of Thermodynamics

The *Third Law of Thermodynamics* states:

That at absolute zero, the entropy is equal to zero for any pure structure.

By raising a pure structure from absolute zero to temperature T, the entropy change is:

$$\Delta S = S_T - S_o$$

or,

$$\Delta S = S_T$$

where,

$S_o = 0$; the entropy of the system at absolute zero.

This equation results in the absolute entropy for the pure structure. The absolute entropy gives a comparison of the amount of disorder that is present in a pure structure and is useful for finding the entropy change. There is a difference in the application of the absolute entropy from a pure structure at a standard-state condition, however. In this case, the absolute entropy will not equal zero.

Therefore, at absolute zero, the absolute entropy for a pure structure is equal to zero. In other words, at absolute zero, all molecular motion within the pure structure has stopped and there is no disorder. However, this statement does not hold true for structures that contain impurities.

THE MAGNETIC FIELD

One of the premier experiments demonstrable in superconductivity is the Meissner effect (see Chapter 5). Basically, this experiment represents the linear association between magnetization and the magnetic field. This linear association dissipates in applications with high magnetic fields. An example of the absence of this linear association is found in a superconductor's critical magnetic field.

For example, if a given superconductor is enclosed in a fixed magnetic field source parallel to its surface, the field distribution within the superconductor will follow the Meissner effect for modest field values. This distribution can be determined by a computation derived from the kernel. As the magnetic field is increased, however, the metal will leave the superconducting state and return to its normal state, exhibiting a latent heat of transition. An interpretation of this phenomenon is that heat must be injected into the metal to force the change of state from a superconducting one back into the normal state. In order to understand this transition, an elaborate application of thermodynamics centering around the free energy change for the system, in terms of the Meissner effect, is necessary.

In a typical superconductor, a magnetic field is emitted from the metal. As this metal returns to its normal state, the magnetic field begins to penetrate the metal. Therefore, the critical magnetic field (H_c) represents an equilibrium between the difference in field energy for both states and the difference in the intrinsic free energy for both states. Due to this difference, there is a significant negative sign that indicates work is being performed or, more importantly, that energy is being supplied to the superconductor.

One method of visualizing the importance of this negative sign is to place a superconductor inside the flux from a magnetic field. In order to actually move the superconductor over the magnet (i.e., the area of greatest magnetic flux), an amount of work is required. A metal in a normal state, however, would not require this work due to the penetration of the magnetic field through the metal. Therefore, the field energy and the free energy must balance at a critical field.

These critical field observations can be proven through a series of thermodynamic equations. Beginning with the stored magnetic work for a given metal:

$$F_m = (4\pi)^{-1} \int_o^B HdB$$

Solving this isothermal situation for the total free energy per cm³:

$$F(B,T) = F_o(T) + (4\pi)^{-1} \int_o^B HdB$$

Based on the Meissner effect:

$$H = B - 4\pi M$$

This results in a thermodynamic potential with reference to H rather than B:

$$G_s(H,T) = F_o(T) - (4\pi)^{-1} \int_o^H B(H)dH$$

Dealing solely with the superconducting magnetic state equation:

$$G_s(H,T) = F_o(T)$$

for,

$$H \leq H_c$$

where,

$$H_c = \text{critical field}$$

As H exceeds the critical field, the superconductor exhibits normal metal magnetic activity:

$$B = H$$

for,

$$H > H_c$$

The equation can now be rewritten as:

$$G_n(H,T) = F_o(T) - (8\pi)^{-1}(H^2 - H_c^2)$$

where the subscript n indicates the metal in its normal state. Therefore, G(H,T) exhibits two separate forms for H. The entropy for this same metal in its superconducting state is:

$$S = -F_o'(T)$$

and in its normal state is:

$$S = -F_o'(T) - (4\pi)^{-1}H_c(dH_c/dT)$$

This results in a latent heat of transition:

$$Q = -(T/4\pi)H_c(dH_c/dT)$$

A careful examination of this last equation indicates that a positive latent heat is necessary for transforming a metal from its superconducting state into its normal state. In application with the previous critical field equations, an adequate explanation would be that the critical field decreases with an increase in temperature. A reasonable interpretation for this feature is that the emission of the Meissner effect is reversible in a superconductor.

GINZBURG-LANDAU THEORY

Building on this insertion of a superconductor into a magnetic field, an entirely different set of phenomena results from an increase in the strength of the surrounding field. Instead of the previously described behavior, the development of a domain structure, with alternating normal and superconducting areas, was theoretically predicted by Landau in the late 1930s and early 1940s. Landau labeled this prediction the *intermediate state of superconductors.* Two researchers, Shalnikov and Meshkovsky, were later able to experimentally verify Landau's intermediate state prediction.

Another feature of Landau's work dealt with the definition of a positive surface energy variable. This energy term was a significant event in the formation of an intermediate "zone" between the superconducting and normal areas. Makei was later (1958) able to demonstrate the relationship of the intermediate state and the magnitude of the surface energy.

When a magnetic field is exercised on a superconductor, there is a complete penetration of the flux through the normal state area while only a partial penetration is possible into the superconducting areas. It is this partial penetration into the superconducting areas that forms the surface energy. In this situation, due to the relationship (linear) between M and B, there is a negative surface free energy.

Abrikosov pointed out that the lowest free energy is derived from a complex organization of "filaments" or flux lines that are found inside the superconductor. These tightly packed flux lines are separated by a distance that is less than a London depth. This spacing allows a magnetic field to penetrate the metal without emitting a Meissner effect (i.e., $H = H_c$). Slowly, over a period of time, a large expanse of these fields begins to emit a Meissner field and the flux lines begin to diminish in their importance. Once the magnetic field reaches a level below the critical field, the Meissner effect is total.

There are certain superconductors, like lead, where this linear explanation fails to work in light of a positive surface free energy. Coupled to this change of sign for the surface energy is an extremely small correction in the field penetration law. The ability to make accurate predictions for both of these phenomena is the basis for the Ginzburg-Landau theory.

The essential ingredient in the Ginzburg-Landau theory is the formation of "superelectrons." These are the electrons that are the major contributors to the formation of the Meissner effect. An assumption of the Ginzburg-Landau theory is that all superelectrons are physically identical with the same coherent wave function. This assumption does not apply to "normal" electrons, however. These electrons, as illustrated by the microscopic theory, follow a Fermi-Dirac distribution with Pauli principle behavior.

An extension of these superelectrons is found in the inducement of persistent currents. This persistent current is experimentally demonstrated through the placement of a metal ring between the poles of a magnet. If this metal ring is cooled into a superconducting state and removed from between the poles of the magnet, then the extraction process forms a current within the ring. This current is called a *persistent ring current*.

Persistent ring currents have been maintained for periods of several years. This longevity leads to the presumption that ring currents can persist indefinitely. Because of the persistent nature of these ring currents, they must be associated with metastable states versus the more contemporary thermodynamic equilibrium states discussed previously in this chapter. This metastable state accounts for a current in the ring to form a magnetic field that establishes a field energy. Thus, this current-induced field energy is far higher than the energy of the currentless ring.

The nature of persistent ring currents supports the assumption that superelectrons maintain their coherent properties over large distances. In this case, the length of the superconductor ring expresses a need for an explanation for the superelectron coherent wave function. Magnetic flux serves as the basis for this explanation. Experiments dealing with the measurement of the magnetic flux inside the "hole" of the superconductor ring found a degree of "quantization" for the flux. Flux quantization serves as verification that a superconductor has global properties.

ISOTOPE EFFECT

One final examination of superconductivity that can be theoretically examined at the thermodynamic level is the Fröhlich theory of electron interaction. As discussed in Chapter 3, electrical resistivity in a metal is dependent on the thermal motion of the crystal lattice structure. Essentially, this motion is dictated by the movement of electrons within the lattice structure. One of the attributes of electron movement is the formation of phonons. In a superconductor, there exists a unique electron-phonon-electron interaction that accounts for either the gain or the loss of electrical resistivity.

Fröhlich stated that this interaction is velocity dependent. As such, the phonons travel at a dramatically slower speed than the electrons. In fact, a ratio of 1:100 is typical between the sonic phonon and the Fermi velocity electron. This speed difference forces a phonon to trail behind the emitting electron. Therefore, other electrons are only able to interact with a phonon (and thus the emitting electron's momentum) by crossing through the trailing wake of the emitting electron.

One assumption Fröhlich made during the formation of his theory for electron-phonon-electron interaction was that normal Fermi distribution for electrons is unstable during Fröhlich interaction. By using this assumption, a lower energy was possible through a higher Fröhlich interaction. Fröhlich used this assumption as a significant means for theoretically distinguishing a superconducting metal from a normal conductor. The Fröhlich interaction theory predicts that:

❖ A superconductor metal is a poor normal conductor.
❖ Lattice motion contributes to superconductivity.

In addition to normal metals, Fröhlich's theory also provides a valid discussion of metal isotopes. This extension, known as the *isotope effect*, declares that the superconducting-state transition temperature is dependent on the isotopic mass of the metal's nucleus. While the Fröhlich interaction theory was derived independently of the isotope effect (Reynolds and Maxwell experimentally discovered the effect in 1950), subsequent verification confirmed the validity of applying Fröhlich interaction to the isotope effect.

The Fröhlich interaction theory is not without its objections, however. Numerous critical literature citations have blasted both the methods and the conclusions derived from this electron-level interaction. Some of the more important arguments include:

✣ Fröhlich's failure to examine the initial onset of superconductivity.
✣ Fröhlich's employment of a perturbation theory.
✣ Failure to adequately describe the Meissner effect.
✣ Too large of an energy difference between the superconducting and normal states.
✣ Ignorance of the Coulomb interaction between electrons.

In order to retain the Fröhlich interaction theory, several researchers composed modifications to the original theory. These amendments were founded in an effort of preserving the integrity of a sound argument for the theory of superconductivity.

5

Superconductor Experiments

There are two frequently demonstrated experiments that amplify the spectacular attributes of superconductivity. Both of these experiments center around the electromagnetic properties of superconductors.

The first of these two experiments reflects the behavior of a superconductor inside a magnetic field. In this experiment, a superconductor is placed between the two poles of a magnet. As the magnetic field nears the superconductor, a field exclusion phenomenon prevents the field from passing through the metal. This blocking of the magnetic field is in direct contrast to the behavior exhibited by the same metal in its normal state; In its normal state, the metal does permit a magnetic field to pass through its structure. Subsequently, cooling this metal to its superconducting state prevents the passage of a magnetic field. This magnetic field exclusion is not total, however. There is a modest amount of field penetration near the surface of the superconductor, approximately 10^{-5} to 10^{-6} cm.

While this magnetic field exclusion experiment is impressive in its relationship to the formulation of a penetration equation (see later in this chapter), another superconductor experiment is more commonly perceived as the definitive illustration of superconductivity. Known as the Meissner effect, or more properly, the Meissner-Ochsenfeld effect, this experiment demonstrates the emission or expulsion of a magnetic field from a superconductor (see Fig. 5-1).

In their now classic experiment, Meissner and Ochsenfeld placed a single crystal of room-temperature tin inside the flux of a magnetic field. This arrangement allowed the field to completely penetrate the tin crystal. They then cooled the tin to its superconducting state. As the tin passed through its superconducting transition temperature, the magnetic field strength increased. Based on this observation, Meissner and Ochsenfeld theorized that the tin must have expelled its own magnetic field as it was cooled to the superconducting temperature. A contemporary modification of this experiment utilizes this magnetic field expulsion as a means for "floating" a magnet over the cooled superconductor (a description of this experiment is presented as a construction project in Appendix A).

Several exciting conclusions can be drawn from this documented expulsion of a magnetic field:

✛ Superconductivity is a magnetic property.
✛ There must be a true thermodynamic state of equilibrium versus an indefinite metastable state.
✛ Superconductivity exhibits a unique form of diamagnetism.

Each of these conclusions can be studied in greater depth through a series of equations and formulas. This mathematical exercise will also shed some light on the nature of the magnetic field expulsion from a superconductor.

A typical magnetic metal in its normal state can be expressed as:

$$B = \mu H$$

where,

$$
\begin{aligned}
B &= \text{density of the magnetic flux} \\
\mu &= \text{a property dependent constant} \\
H &= \text{intensity of the magnetic field}
\end{aligned}
$$

In this representation, B is considered to be a more discrete quantity than H.

Relating a magnetization density variable, M, to this normal magnetic state results in:

$$H = B - 4\pi M$$

(Photograph courtesy of AT&T Bell Laboratories.)

Fig. 5-1. A demonstration of the Meissner effect.

As a point of reference, M = 0 in a total vacuum. Therefore, M serves as a suitable measurement for the response of the metal to the surrounding field. This relationship is better illustrated by:

$$M = \chi H$$

where,

$$\chi = \text{magnetic susceptibility}; \chi = (\mu - 1)/4\,\pi$$

In order to properly express H, an important state modification must be made to the original equation. This results in the basic magnetism equation:

$$M = KB$$

where,

$$K = \text{constant related to both } \mu \text{ and } \chi; K = \chi/\mu$$

While this equation is an excellent description of a normal magnetic metal, it fails to accommodate either ferromagnets or superconductors. A more accurate description is through the use of supercurrent densities, j_s:

$$\text{curl } (\Lambda j_s) = -B/c$$

where,

$$\Lambda = \text{constant}$$
$$c = \text{speed of light}$$

Metals with an unequal magnetization density (M(r)) require a rewriting of the supercurrent density as:

$$j_s = c \text{ curl } M$$

Now, by inserting this new value for the supercurrent density into the above equation, a better relationship can be shown as:

$$\text{curl curl } M = -B/\Lambda c^2$$

This equation is referred to as the *London equation.*

One use of the London equation is in accurately depicting the expulsion of a magnetic field. In this case, a superconductor is inserted into a magnetic field B_0 which is parallel to the metal's z-axis. This theoretical superconductor will have its boundary points at $-a$ and a. From this example, both M and B are restricted to z-components. Therefore,

$$d^2M/dx^2 = B(x)/\Lambda c^2$$

Solving for $B(x)$ requires the use of the Maxwell equation:

$$c \text{ curl } H = 4 \pi j_{ex} + \gamma D/\gamma t$$

where,

$$j_{ex} = \text{external current density}$$
$$D = \text{electric displacement vector}$$

Because this is a static situation, the result is:

$$dH/dx = 0$$

Plugging this value into the original magnetization density equation:

$$H(x) = B(x) - 4 \pi M(x)$$

Based on the solution of this equation in terms of $B(x)$, the theoretical superconductor equation becomes:

$$d^2B/dx^2 = 4 \pi B/\Lambda c^2$$

The squared length dimension of the righthand denominator allows the insertion of the London penetration depth λ ($\lambda = c(\Lambda/4 \pi)^{1/2}$) into $B(x)$ with a boundary of $B(a) = B(-a) = B_0$:

$$B(x) = B_0(\cosh (x/\lambda)/\cosh(a/\lambda))$$

This result is called the *penetration law for superconductors*. This same penetration law as applied to a normal metal is:

$$B(x) = B_0/(1 - 4 \pi K)$$

A conclusion that can be drawn from this penetration law is that the magnetic field expulsion from a superconductor is nearly total for London penetration depths of 10^{-5} to 10^{-6} cm. Furthermore, the magnetic field near the center of the superconductor (B(0)) is an exponentially small value. Therefore, both of these features contribute to the Meissner effect.

From these two penetration equations (superconductor & normal metal), the unique form of diamagnetism found in superconductors can be shown. A typical diamagnetic metal has a negative constant "K." This sign results in B(x) being less than the magnetic field B_0 in a normal diamagnetic metal. Conversely, B(x) is significantly related to the depth of penetration in a superconductor. This difference in relationship of B(x) to the depth of penetration is indicative of a superconductor's unique form of diamagnetism.

Protection of the interior of the superconductor from the external magnetic field is supplied by the supercurrent density. In a superconductor, the supercurrent density is highest near the surface of the metal. This concentration of the supercurrent density acts as a shield against the applied magnetic field. Contrary to an initial presumption, this concentration is not equivalent to a mathematical surface current. Generally speaking, a mathematical surface current lacks application to a physical sample, such as a superconductor.

The unphysical nature of a normal diamagnetic metal's penetration law can be explained as a matter of error regarding ring currents. These ring currents, as found in a normal diamagnet, have a diameter equivalent to the de Broglie wavelength of an electron at the metal's Fermi surface. In a typical diamagnetic metal, these ring currents are confined to a narrow area. Failure to observe the size of the ring currents gives rise to the false assumption that this is a mathematical surface current phenomenon.

One theorem that is derived from the influence of magnetization density states *that current density that is resultant from magnetization density yields a zero net charge transport.* Based on this statement, both the Meissner effect and charge transport are separate and distinct attributes of a superconductor.

Backtracking to the London equation, experimental evidence supports the bulk of the conclusions that can be drawn from this relationship. Modifications to this equation are necessary, however, for answering all of the phenomena that have been reported for the Meissner effect. This amplification of the London equation begins with a symmetrical superconductor ($-a < x < a$ and $B = B_0$) where the Fourier expansion of the magnetization is:

$$M(x) = \sum_{n=14}^{\infty} M_n \left(a^{-\frac{1}{2}} \cos \left[n - \frac{1}{2} \pi \, x/a \right] \right)$$

or, in a simpler form:

$$M(x) = \sum_{n=1}^{\infty} M_n \, u_n \, (x)$$

Extending this equation with conditions that are acceptable for any example requires the insertion of the vector field $u_n(r)$. Due to the presence of this new vector field, $u_n(r)$ can be interpreted as an eigenfunction of the *curl curl* operator in the London equation:

$$\text{curl curl } u_n = q_n^2 \, u_n$$
$$\text{div } u_n = 0$$

where,

$$u_n = \text{normal to boundary surface of V}$$

$$M(r) = \sum_{n=1}^{\infty} M_n \, u_n(r)$$

The curl curl eigenfunction equation is controlled by the values of the two derivative equations. In this condition, the eigenvalue is limited to positive eigenvalues q_n^2. Therefore, div $u_n = 0$ becomes a condition of the vector function u_n.

Function u_n is a full set for B when div B = 0, therefore, the flux density vector B(r) can be written in terms of u_n:

$$B(r) = \sum_{n=1}^{\infty} B_n \, u_n(r)$$

or, for $u_n(x)$:

$$B(x) = \sum_{n=1}^{\infty} B_n \, u_n(x)$$

Both of these equations offer expansions of the parallel B. By combining these expansions with the London equation, extensive differentiation results in:

$$M_n = K(q_n) \, B_n$$

In this expression, the term K(q) is called the London kernel. This London kernel is both a Fourier transformation of K(r,r') and a generalized form of the constant K (from the original magnetic metal equation). In a superconductor, the London kernel can be represented as:

$$K(q) = -1/\Lambda c^2 q^2$$

While in a normal metal, a similar kernel would be viewed as:

$$K(g) = \grave{}K/(1 + (\mu q)^2)$$

The difference between these two kernel equations stems from the behavior of $K(q)$ in reference to the wave number (q). For a normal magnetic metal, $K(q) = K$, no matter what the magnitude of q. Therefore, $K(q)$ establishes the field penetration law for the surface of the metal when a large q is present; there is little regard for whether the metal is a superconductor or a normal state sample. In terms of small q values, however, there is a significant difference in the behavior of these two kernels.

A small q results in the normal metal kernel becoming finite, while the superconductor's kernel becomes infinite. Conversely, $K(q)$ determines the field penetration law for distances deep inside the magnetic metal. The conclusion that can be drawn from this observation is that the field expulsion is related to the kernel $K(q)$ where there is a $1/q^2$ singularity at $q = 0$. Through experimental verification, this point has become a major theorem in distinguishing superconductors from normal metals.

While $K(q)$ at $q = 0$ is an important advancement in understanding the magnetic behavior of a superconductor, Pippard in 1953 proposed some experimental modifications to $K(q)$ at intermediate values of q. Essentially, Pippard derived a vital length in superconductors that supplemented the London penetration depth. He used three variables in the formulation of his aptly named *coherence length factor*:

✧ Surface energy magnitude.
✧ Transition definition.
✧ Impurity content field penetration.

Pippard's coherence length definition was derived from his belief that this new distance factor was based on a correlated wave function.

Pippard discovered that the largest influence on the coherence length was the impurity content of the superconductor. Oddly enough, the coherence length was found to *decrease* with the addition of impurities.

In difference to this reduction, a pure superconductor was found to exhibit a coherence length of 10^{-4} cm. This value is significantly larger than the previously quoted London depth of 10^{-5} to 10^{-6} cm. Based on this phenomenon, the kernel can be rewritten as:

$$K(q) = -(1/\Lambda c^2 q^2)(\arctan (\xi q))/\xi_o q$$

where,

ξ = coherence length
ξ_o = coherence length limit due to impurity

Interestingly enough, while this "Pippard kernel" offers viable values for intermediate q, there is very little difference between the London kernel and this Pippard kernel with relation to small q. Therefore, in terms of x, the Pippard kernel offers a unique field penetration law for small x, but provides a similar exponential decrease as the London field penetration law for large x.

An area where the London kernel does provide greater insight into the composition of a superconductor is in the definition of a "perfect" superconductor. M. R. Schafroth demonstrated such a superconductor theoretically in 1955. Basically, Schafroth's perfect superconductor is a magnetic sample with a kernel $K(q)$ and a yq^2 singularity at $q = 0$, where there is a total field expulsion for large x. Schafroth explained further that a perfect superconductor is similar to an equilibrium superfluid. This similarity is found in the spinless particles of a superfluid and the same particles undergoing a charge in a superconductor.

An extension of Schafroth's perfect superconductor theory stresses the significance of the moment of inertia under rotation as being a major contributor to superconductivity. Both of these relationships form the basis of a connection theorem for superconductors and superfluids. Taken in this format, the perfect superconductor becomes a derivative of quantum mechanics.

Actually, this quantum mechanical correlation can only be actively proved through the interaction of both the previously discussed microscopic theory (see Chapter 3) and superconductor thermodynamics (see Chapter 4). Only through the implementation of these two disciplines can a reasonable explanation for the application of a quantum mechanical correlation in a classical system over an arbitrarily long distance be expressed. Unfortunately, failure to readily accept this implementation has hampered the widespread acknowledgment of the preceding phenomenological equations for the Meissner effect.

Any discussion of the Meissner effect cannot end without a modest treatment of Schafroth's "pseudo-molecules." In a short paper presented in 1958, Schafroth offered two explanations for the Meissner effect:

❖ Beginning with $H = H_{electrons} + H_{phonons} + H_{interacton}$, this theory used canonical transformations and ignorance of certain parameters. All of these steps must be carefully balanced for a current density in a fixed magnetic field. Schafroth dismissed this theory as unsuccessful.

❖ The theory endorsed by Schafroth used pseudo-molecules that featured an undisturbed center-of-gravity motion. Consequently, only the internal wave function of these pseudo-molecules is affected by the electron-phonon interaction. This concentrated effect leaves the center-of-gravity motion virtually free. Furthermore, these pseudo-molecules must exhibit an adherence to Bose statistics. In Schafroth's conclusion, this theory would be ample for providing the Meissner effect.

6

Real-World Superconductor Applications

Only after all of the theoretical examinations of superconductors have been concluded can the practical applications of this technology be realized. In fact, understanding the mechanics of *how* a superconductor works is far simpler than deciding *what* to do with superconductivity.

Although many superconductor applications are currently being explored, the two areas that show the greatest practical promise are superconducting wires and superconducting magnets. Furthermore, these are the two application areas that have generated the greatest amount of research and development.

SUPERCONDUCTING WIRE

Early attempts at drawing a superconductor into a wire form were fraught with disappointment. Generally, the delicate and brittle nature of the superconductor prevented the drawing of the metal into a wire (see Fig. 6-1). Eventually, a special cable arrangement was developed for combining superconductor filaments with copper support fibers. There were two typical arrangements for this composite superconducting cable. One cable type featured a single superconducting wire surrounded

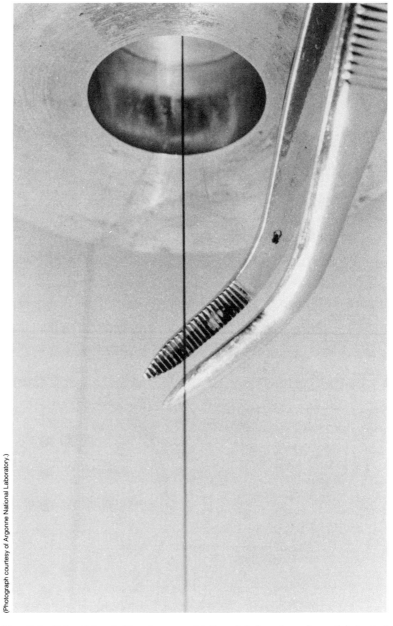

Fig. 6-1. Scientists at the Argonne National Laboratory have fabricated superconducting ceramics into a wire shape.

by a pure copper sheath. The other cable arrangement consisted of a multi-strand superconductor array supported inside a pure copper field (see Fig. 6-2).

Both of these cable configurations are considered stable superconducting cables. The single wire form is suitable for low-field usage, and the multi-strand cable is useful for high-field applications. Three of the more common superconductors used in cable preparation are alloys of niobium: niobium-tin, niobium-titanium, and niobium-zirconium. Of these three, the niobium-tin alloy proves to be the most difficult to draw into a wire. The niobium-titanium and niobium-zirconium alloys are malleable enough for direct use in the wire formation process. Niobium-tin's fragility, on the other hand, mandates that this alloy be housed within a substrate prior to wire fabrication.

Fixing the niobium-tin alloy to a wire is generally accomplished through a deposition method. This chemical process involves the reduction of niobium (gaseous) and tin chloride into a vapor alloy that is deposited on a stainless-steel wire substrate. Alternatively, a similar deposition can be produced through the diffusion of tin onto a niobium wire.

The process for making a niobium-titanium or a niobium-zirconium wire differs significantly from the deposition methods used in the

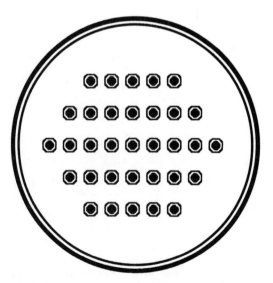

Fig. 6-2. A simple copper-matrix superconducting cable arrangement. The dark-colored superconducting filaments are held within a copper substrate.

production of a niobium-tin wire. Unlike other superconductors, these two niobium alloys must be devoid of impurities. In order to achieve this high level of purity, the niobium and its alloy are subjected to arc-beam melting techniques. Eventually, the final wire is exposed to a concluding heat bath.

Once the fabrication technique has been determined, there are two physical features inherent in any superconducting cable application:

- ✤ **mechanical strength**—the ability of the superconducting cable to withstand enormous internal and external stress.
- ✤ **stabilization**—the predictable and stable behavior of the superconducting cable under normal operating conditions.

Mechanical Strength. Regardless of the alloy or fabrication technique used in the construction of the superconducting cable, there will be considerable internal and external force exerted on the cable. Failure to adequately compensate for these forces will result in the complete loss of the integrity of the superconductor.

The need for mechanical strength is most evident in the use of superconducting cables for constructing high-field magnets. In this case, the force that is exerted on the cable is derived from a current and magnetic-field Lorentz force. The destructive nature of this force is checked by the tension in the cable windings on the magnet's coil. Another detrimental force is the axial compression force on the turns of the coil.

These counterproductive forces can exert outward pressure tensions greater than 600 pounds per square centimeter in an average-sized magnet. These pressure extremes result in the reversion of the superconductor to its normal state. Therefore, the strengthening of the superconducting cable through copper sheaths or substrates is imperative for proper magnet operation.

Coil construction is also useful in combating the negating effects of axial compression. Split coils offer an excellent example of this requirement. In a split coil, axial compression serves to pull the two subcoils together. This effect results in a cumulative force that is greater than that found in a similar single coil magnet. Only stabilized cables and unique coil winding patterns can prevent the destruction of the magnet's superconducting portion.

One final area where mechanical strength is vital in the construction of a superconducting magnet is in the cooling system design. In order to be efficient, the superconducting cable must be maintained below its critical temperature. Obtaining this temperature level with niobium alloy cables demands the use of a liquid helium coolant circulation system. Introducing this cooling system into a magnet can only be accomplished through using stabilized cables and minimizing the current density of the magnet. The construction of high-field magnets, however, requires an alternate design technique (see later in this chapter).

Stabilization. A stabilized superconducting cable facilitates the construction of low-temperature magnets. Another attribute of a stabilized cable is its ability to carry elevated current densities that are higher than those obtainable in non-stabilized superconducting cables.

A typical niobium-alloy stabilized cable uses copper as a stabilizing agent. In this function, the copper serves as a thermal conductor between the liquid helium coolant and the internal superconductor. Under normal operating conditions, as the current density forces the superconductor to move towards its normal state, the heat increase is localized by the surrounding copper matrix. The copper then acts as a thermal conductor between the liquid helium bath and the overheated superconductor for cooling the niobium alloy back into its superconducting state.

The copper matrix also has the ability to serve as a resistance shunt between the normal and superconducting states. In this example, the copper forms a "bridge" from the superconducting-state niobium alloy and the normal-state niobium region. This link is used for returning the current flow to a superconducting condition once the superconductor has been returned to its operational state.

There is a limit, however, on the current density that even a stabilized cable can accept. If the current goes too high, for example, the surrounding liquid helium bath will exhibit film boiling. Film boiling is an overly aggressive heating of liquid helium above its normal nucleate boiling level. The best means of reducing the affect of film boiling is through fully stabilized cables. A fully stabilized cable can maintain its integrity up to the limit for the critical current/liquid helium temperature factor.

Fully stabilized superconducting cables are constructed in numerous configurations. This varied shape and size is determined by the design of the liquid helium coolant system, as well as the intended winding of the magnet's coil (see Fig. 6-3). In addition to this physical appearance, experimentation has been performed on the relationship between the torque or pitch of the cable's winding and the suppression of adverse

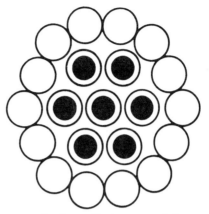

***Fig. 6-3. A stable superconducting cable features solid copper wires wrapped
around several copper-matrix superconducting cables.***

current densities. The preliminary evidence from these experiments
indicates that the stability of a superconducting cable can be enhanced
through filament twistings, transpositions, and copper-alloy matrix
substitutions. The success of these modifications has paved the way for
the design of high-field magnets.

HIGH-FIELD MAGNETS

A conventional high-field magnet producing 30k Oe (oersteds) is
designed with a water-cooled copper coil that receives a large current.
Unfortunately, the power requirements for this conventional magnet are
costly. For example, to generate 100k Oe in a modest magnet would
require the application of over one megawatt of power. Furthermore,
a large portion of this power input is lost as joule heat. This large degree of inefficiency makes the design of a superconducting magnet
desirable.

Only one power requirement is needed by a superconducting
magnet—operating its cooling system. In other words, once the magnetic
field has been established, the magnet can be placed in a *persistent
mode*. A persistent mode is formed when the coil's wire ends are fixed
to a superconducting link and the power supply is removed. Therefore,
the only continuous power requirement for a superconducting magnet
is in the maintenance of its cooling system.

A fringe benefit of operating a superconducting magnet in a persistent mode is the elimination of external noise. This noise elimination
is triggered by a steady region of magnetic flux inside the coil due to

a superconducting ring formed from the coil and superconducting link configuration.

High-field magnet construction was originally undertaken with niobium wire (by Yntema in 1955). In these early attempts, a magnet with a field of 5k Oe was constructed. Further advances were made when niobium alloys (a team lead by Kunzler in 1961) were substituted for the niobium wire. These electromagnets were able to maintain a field of 88k Oe (see Fig. 6-4). In spite of these accomplishments in magnet construction, two problems still plague magnet designs:

❖ wire length
❖ coolant system

Wire Length. There are three factors that are in conflict when specifying the amount of wire needed in an electromagnet design: maximum field production, homogeneity, and cost. Attempts at solving these three pitfalls in magnet design center around specialized wiring patterns. The more successful wiring arrangements include: rectangular cross-sectional solenoids (Boom and Livingston in 1962), Helmholtz pair coil construction (Day in 1963), and dual design curves featuring the minimal amount of wire (by Thomas and Bright in 1966).

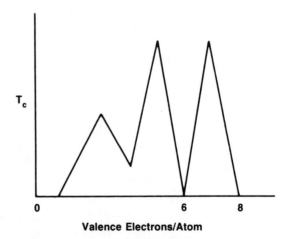

Fig. 6-4. In high-field magnets, Matthias (1957) showed that the superconductors with the highest critical temperature were those with a 3, 5, or 7 valence electron/atom ratio.

Coolant System. Virtually, all of the power required to maintain a superconductor magnet is used to operate a cooling system. The efficiency of this cryostat is dictated by a gauge dilemma in the superconducting wire. In order to minimize the amount of outside heat influence, a small gauge wire must be used throughout the magnet design. Contrary to this small gauge, in order to reduce the amount of joule heat, a large gauge is needed. Therefore, a dilemma existed in choosing the optimal wire gauge for winding a solenoid. A team of researchers lead by Goedemoed in 1965 developed a simple solution to this gauge dilemma. A flux pump sends small packs of flux into the superconducting solenoid. This flux transmission is activated by a revolving magnet that forces selected areas of the solenoid into a normal state.

One undesirable effect from this simple flux pump arrangement is that the rotating magnet generates heat inside the cryostat. Therefore, other methods of injecting flux into the solenoid have been developed. Two of the more popular flux injection methods have been through thermally switched circuits and flux pattern movement. Of these two methods, only the flux pattern movement technique offers the lowest amount of external heat introduction. In this case, the flux movement is initiated through an externally mounted ac motor.

Unfortunately, all of the problems centering around superconducting magnets have not yet been solved. Flux trapping, pinning points, and remanent fields all reduce the efficiency of these electromagnets. In spite of these limitations, fields of 100k Oe are commonly produced with today's high-field superconducting magnets.

OTHER APPLICATIONS

While many researchers are wildly speculating about the potential for superconductor applications (see Chapter 1), greater experimentation must be conducted prior to realizing these dreams. The leading candidate for satisfying these real-world application goals is the high-temperature superconductor.

All of the preceeding applications discussion has dealt with low-temperature superconductor research. With the introduction of newer, more stable, high-temperature superconductors, however, other applications will become possible. Two of the more probable application areas for receiving this renewed interest will be:

✦ magnetic bearing
✦ ac transformers

Magnetic Bearing. Building a magnetic bearing from a superconducting cylinder and disk was performed by Buchhold in 1960 (see Fig. 6-5). This bearing design exercised the flux trapping and flux exclusion possibilities demonstrated in some superconducting magnets. Buchhold's bearing was formed from two superconductor pieces (the cylinder and the disk) with very little friction from outside influences. In this design, the disk was made to float on the flux with its resultant spin controlled through the spin of the cylinder. Costly and complicated cooling systems reduced additional research in this application area. Only through the introduction of efficient high-temperature superconductors will the design of a superconducting magnetic bearing become feasible.

Ac Transformers. An ac transformer (or any heavy electrical engineering machine) built from superconducting cable results in a significant reduction in the size, weight, and cost of the central coil. In fact, Wilkinson (in 1963) predicted that the proper superconductor

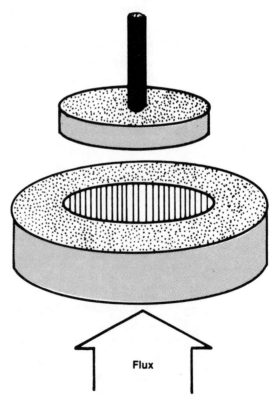

Fig. 6-5. A superconducting bearing based on Buchhold's 1960 design.

arrangement would result in a 40 percent reduction in these three factors. Much of this savings is due to the elimination of extensive copper wiring from the superconductor-based coil. Once again, the limitation of coupling the superconductor coil with an extensive cooling system has restricted the development of superconductor power transformers. High-temperature superconductors aren't the single answer in the realization of this application, however.

Even though high-temperature superconductors require a reduced cooling system, they still aren't cost-effective. In this case, a liquid nitrogen coolant is circulated through the system instead of the liquid helium bath used in low-temperature superconductors. While the cost of liquid nitrogen is tenfold lower than liquid helium, the cost of installing this cooling system is still higher than the cost of building a normal transformer. Furthermore, the yield from the superconductor transformer is not comparable to that obtained from a similar conventional copper-based transformer. Therefore, the bottom line in superconductor applications is linked to the solution of three key factors:

✥ Low-cost superconductor manufacturing
✥ Cooling system elimination
✥ Efficiency in superconductor operation

Only after these three factors have been satisfactorily addressed can the development of significant real-world superconductor applications begin.

Appendices

Appendices

Appendix A

Building a Superconductor Project

Turning superconductivity theory into a superconductor project is a relatively simple recipe thanks to the advances in high-temperature superconductors. Based on a chemical combination of three metal oxides, this ceramic perovskite superconductor can be "cooked up" in virtually any modest laboratory environment. Once the superconductor has been properly prepared, only a cooling bath of liquid nitrogen will be necessary for transforming this normal metal into a superconducting one.

WARNING

Before attempting any of the following experimental procedures, make sure that your laboratory environment is properly prepared for handling the compounds required in the formulation of superconductors. If you are unsure of these requirements, consult with your local university-level science department.

Up until late 1987, there was only one generally accepted method for manufacturing a high-temperature perovskite superconductor. This technique was perfected by the founding fathers of high-temperature superconductivity—Müller and Bednorz (1986). Their manufacturing process began with a compound of copper oxide (later Chu, in 1987, would modify the high-temperature superconductor formula by adding yttrium oxide and barium oxide, a derivative of barium carbonate, and copper oxide). The resultant oxide compound is then combined with citric acid and ethylene glycol (piston-engine radiator coolant). This mixture is finally heated to a temperature of 100 degrees Fahrenheit.

After this mixture has been heated, the remaining liquids are vaporized by baking the oxide combination in a 1500-degree Fahrenheit oven. A fine black powder is the final residue from this baking step. While the powder is still hot, it is placed inside a press and compressed at 3000 psi. Lastly, this resulting wafer is slowly cooled over an 8-hour period. *Voilà*—an instant high-temperature superconductor.

Even though Müller and Bednorz had their fair share of skeptics, the result of this "homebrew" experiment was a 94-degrees Kelvin superconductor. Independent laboratories were later able to verify the existence of superconducting properties that were attributed to this compound by the two IBM research scientists. In recognition of their landmark achievement, Müller and Bednorz were awarded a Nobel Prize in 1987.

MAKING A SUPERCONDUCTOR

Once Müller and Bednorz had laid the groundwork for high-temperature superconductors, other researchers improved on their formula. Teams like the one lead by Paul Chu at the University of Houston continued to march the superconductor temperature up the Kelvin scale. Finally, another IBM research scientist, P. Grant (at the IBM Almaden Research Center), discovered a simplified formula for a yttrium, barium, and copper oxide superconductor (see Fig. A-1). It is Grant's formula that makes superconductor construction projects accessible to virtually any laboratory (see Fig. A-2).

In just seven simple steps, a superconductor can be formulated from a handful of simple, laboratory-grade chemicals. Once this chemical combination has been formulated, two common laboratory apparatuses are required: a furnace and a press. The following step-by-step procedure offers suggestions on how the potential limitation from each of these requirements can be minimized.

Making an Oven. In order to be effective, an oven used for creating superconductors must be able to generate temperatures in the neighborhood of 900 to 1000°C. There are many laboratory-grade furnaces that can meet this temperature requirement, but unfortunately,

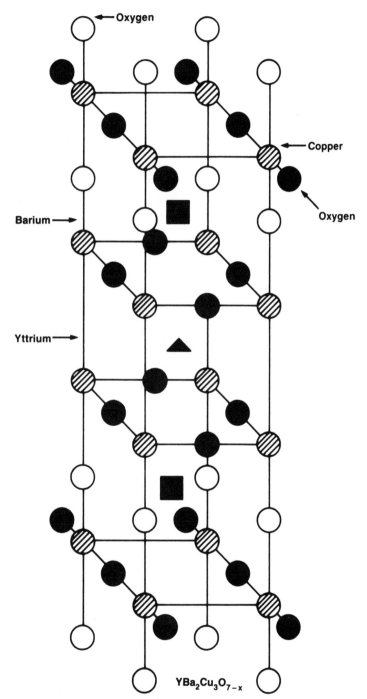

Fig. A-1. An ideal high-temperature superconductor recipe.

(Photograph courtesy of IBM Research Division.)

Fig. A-2. A scanning electron micrograph (SEM) of a single "1-2-3" crystal.

these expensive devices are beyond the budgets of most small laboratories.

An alternative to buying a sophisticated professional furnace is to use a common sculptor's kiln. A high-temperature kiln can either be purchased, or more importantly, leased from many art departments at universities and colleges.

During the preparation of a superconductor, two consecutive days of kiln time will be necessary. Furthermore, the superconductor *must* be the only occupant of the kiln while it is baking or curing. These two important points should be discussed with the art department chairman when negotiating the lease of kiln time.

Making a Press. The only other piece of laboratory equipment required for preparing superconductors is a press. Once again, the specifications for a professional press are beyond the means of the average superconductivity experimenter. A press for compressing superconductor powder into superconductor disks must be able to produce 7 to 9 tons psi.

There is a low-cost alternative to purchasing this device from a laboratory supply house. Presses capable of compressing powders to the required pressure range can be found at local machine shops. This work is inexpensive and only requires the purchase of a small disk die for holding the superconducting powder prior to its compression.

Additional sources for large hydraulic presses are high schools, colleges, and universities. As with the machine shop press, a small disk die is needed for compressing the superconducting powder into a superconductor disk. In both cases, the die should be relatively small—½ to 1 inch in diameter. Your die selection will also be dictated by the size of the press's anvil. Consult with the press operator as to die and anvil requirements.

MIXING, SHAPING, AND HANDLING SUPERCONDUCTORS

Once an oven and a press have been secured, the remaining steps in the preparation of a superconductor are as easy as following a conventional cake recipe. There are three individual steps used in making a superconductor: mixing the chemicals, shaping the superconductor disk, and handling the superconductor disk. The following outline covers each of these points in greater detail.

Mixing the Chemicals. During this step, you need the following items:

✛ a mortar and pestle
✛ a small spatula or scoop
✛ a scale; 0.1 gram sensitivity
✛ an oven, kiln, or furnace
✛ a crucible; must be bake-resistant
✛ 1.1 grams yttrium oxide
✛ 3.9 grams barium carbonate
✛ 2.4 grams copper oxide

Procedure:

Step 1: Measure each of the chemicals into the mortar.

Step 2: Grind the three chemicals into a fine powder.

Step 3: Place the powder in the crucible.

Step 4: Bake the powder in the furnace for 12 hours at a temperature of 950°C.

Step 5: Cool the powder inside the furnace for 6 hours.

Shaping the Superconductor. You need the following items for this step:

✛ a mortar and pestle
✛ a small ½- to 1-inch die
✛ a 9-ton psi press
✛ a furnace

Procedure:

Step 1: Pour the baked powder into the mortar and grind the mixture.

Step 2: Place the ground powder in the die.

Step 3: Slip the anvil inside the die and on top of the powder.

Step 4: Compress the powder.

Step 5: Bake the disk in the furnace for 12 hours at 950°C.

Step 6: Cool the disk *slowly* inside the furnace for 8 hours.

Handling the Superconductor Disk. The following items are required for performing this step:

✛ plastic tweezers
✛ a glass stender dish w/lid
✛ the superconductor disk

Procedure:

Step 1: Handle the superconductor disk *only* with plastic tweezers.

Step 2: Carefully place the disk inside the stender dish and secure the lid.

Step 3: Avoid sudden movements of the stender dish.

This newly constructed superconductor disk can now be used in a simple confirmation experiment. This experiment will verify that the disk is truly superconducting.

PRODUCING THE MEISSNER EFFECT

In order to place the disk in its superconducting state, the following materials are needed (see Fig. A-3):

- ✛ a superconductor
- ✛ one 300 ml glass beaker
- ✛ glass culture dish
- ✛ 500 ml of liquid nitrogen
- ✛ plastic tweezers
- ✛ a 1-inch diameter rare earth cobalt magnet disk

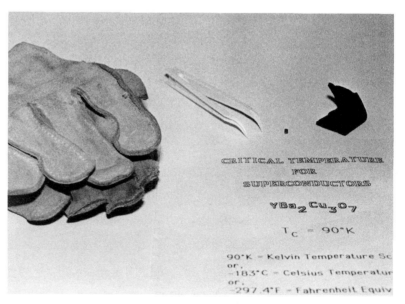

Fig. A-3. Very few materials are needed for demonstrating the Meissner effect.

Procedure:

Step 1: Gently place the superconductor disk inside the glass beaker with the plastic tweezers.

Step 2: Set the magnet in the culture dish.

Step 3: Place the beaker on top of magnet.

Step 4: Pour enough liquid nitrogen into the culture dish to submerge the magnet and ring the beaker's bottom.

As the superconductor disk cools, it will begin to float above the magnet. The beaker is used to restrict the sideways movement of the disk. In certain room temperature situations, the beaker might insulate the superconductor from receiving the complete cooling effects of the liquid nitrogen, and the disk might fail to float above the magnet. As possible solutions to this situation, try the following:

Option 1. Switch positions with the magnet and the superconductor disk. Using this option, however, requires the use of a smaller (lighter) magnet—slice a thin portion of the rare earth magnet off and place it in the beaker.

Fig. A-4. Carefully pour the liquid nitrogen into a shallow dish. An insulated coffee cup is a suitable substitute for the more conventional glass culture dish.

Option 2. Place a small slice of the rare earth magnet directly on top of the superconductor disk. (The beaker will not be needed with this option.) Immerse the superconductor disk in a pool of liquid nitrogen. Once the superconducting state is reached, the magnet slice should float above the disk (see Figs. A-4 through A-9).

Preparing your own superconductors can be a costly adventure that must be confined to a clean laboratory environment. As an alternative to the actual fabrication process, two companies offer ready-made superconductors for consumer purchase. Colorado Superconductor, Inc. and Furuuchi Chemical Corporation each market complete superconductor experimenter's kits.

The domestically available Colorado Superconductor kit is an inexpensive offering. Named the Complete Demonstration Kit, this product features a 1-inch diameter superconductor disk, a rare earth cobalt magnet, and a pair of plastic tweezers. In addition to this "hardware," there is a thorough set of documentation for guiding the beginning experimenter through all of the steps necessary to producing the Meissner effect (see Fig. A-10). One added benefit of the Complete Demonstration Kit is the inclusion of overhead projector transparencies. These materials provide the purchaser with the ability to demonstrate the Meissner effect

Fig. A-5. Allow the liquid nitrogen to reach a stable temperature. Note: the bottom of a wax-coated soft drink cup is being used successfully in this demonstration.

Fig. A-6. Once a stable temperature has been reached, gently lower the superconducting disk into the liquid nitrogen bath with plastic tweezers.

to large audiences. This is an exciting plus for schools and traveling guest lecturers.

Colorado Superconductor also markets three other kits. These kits differ from the Complete Demonstration Kit either in their sophistication (the Advanced Demonstration Kit adds a thermocouple to the basic kit) and in their number of supplies (the Lab Kit for Six Groups includes enough materials for six simultaneous Meissner effect demonstrations). Along with their increase in sophistication and supplies, these additional kits also sport higher prices.

An international flavor in superconductivity experimentation is available from Furuuchi Chemical Corporation. Collectively called the UFO FINE-90 Fantastic Superconduction Kit, there are three kits, each of which contain all of the materials needed for demonstrating the

Fig. A-7. Magnetic levitation through the Meissner effect. This levitation is short-lived, approximately 5 seconds in a room-temperature wax-coated soft drink cup.

Fig. A-8. A dense covering of "ice" forms on both the rare earth magnet and the superconducting disk following the dissipation of the liquid nitrogen.

Fig. A-9. This booth, set up at the 1988 American Booksellers Association Convention and Trade Exhibit in Anaheim, CA was used to perform a demonstration of the Meissner effect. An overhead projector was turned on its side for displaying the final magnetic levitation effect to the audience (author pictured).

Meissner effect. Labeled Set A, Set B, and Set C, each kit includes additional support items for safely dealing with liquid nitrogen. For example, all three kits include one superconductor, five magnets, a set of laboratory dishes, and a plastic tweezer. These supplies constitute the basic Set A kit. Building on this basic kit, Set B adds a metal experiment dish, leather gloves, and a pair of goggles. Finally, the ultimate kit, Set C, contains all of the above items plus a 5-liter metal vacuum jar. This jar is used for holding and transporting liquid nitrogen.

By and large, either one of these manufacturers should be consulted prior to attempting any experimentation with superconductors. Their sophisticated resources allow for the production of superior grade superconductors. This higher quality could spell the difference between a bubbling chemical mess and a successful superconducting demonstration.

Fig. A-10. The Meissner effect displaying magentic levitation.

Appendix B

Elementary Magnetism

Magnetism is an invisible force that defies an easy explanation. An early attempt at forming a theory of magnetism was voiced by André-Marie Ampère. Ampère believed that the magnetism of a material was determined by the orientation of molecules and the molecules which contained circulating currents. Even though Ampère's theory was weak in its exactitude, it does provide a solid foundation for today's theory of magnetism.

Building on this basic beginning, a theory of magnetism starts at the atomic level of matter. Essentially, matter is composed of three successively more complex particles; an *electron* is the smallest and the *domain* is the largest. Each of these particles acts as a subdivision of matter. The electrical nature of magnetism makes the electron the first significant player in the formation of a theory.

A simple explanation of magnetism at the electron subdivision of matter can be demonstrated with the following example. A magnetic field is established by any electron (q) moving at a velocity V to a point P. The resulting magnetic field is determined at P based on the direction of V and the relationship between the plane of V versus the plane of P.

Applying this magnetic field formation concept to electrons orbiting a nucleus, each electron produces an *angular momentum* or *moment of momentum.* Subsequently, these electrons also have a *magnetic moment.* Based on the preceding example, these two moments are in opposite directions for the orbiting electron.

The next higher subdivision of matter is the atom. A typical atom consists of a spinning proton and neutron mass called a nucleus that is ringed with orbiting electrons. Finding the magnetic moment for an atom requires the magnetic moment of the positive-charge axial spin, the magnetic moment of the negative-charge axial spin, and the magnetic moment of the negative-charge orbital spin.

In molecules, the magnetic moment of the electrons is often neutralized. This neutralization results in the proton spins revolving in opposition (para-) or parallel proton spin movement (ortho-) molecules. Contrary to the casual observation of these directions of proton movement, both para and ortho molecules exhibit roughly the same magnetic moment. This equality is derived from the precession of the rotating charges. No matter which manner of orbital movement the molecule is expressing, the charge precession establishes a field that is opposite to the applied field. Therefore, a para molecule will demonstrate a reduction in its magnetic field based on the precessional magnetic moments. This condition is called *diamagnetism.* Noble gas atoms and certain ions (Na and Cl) are examples of diamagnetic materials.

If an electron (or electrons) is removed from a molecule, the result is a singly charged ion. This ion lacks an equality in spin and orbital magnetic moments. Therefore, the ion aligns its orbital and spin magnetic moments in the direction of the applied field. This reinforced field condition is called *paramagnetism.*

Removing the external field from either a diamagnetic or a paramagnetic material voids the magnetic effects in both of these materials. This magnetic effect loss in a diamagnetic is due to the elimination of precessional movements. Similarly, a paramagnetic will lose its magnetic effects due to the equality in its magnetic moments (i.e., the sum is zero).

A group of magnetic materials includes the *ferromagnetics.* These materials enable the production of enormous flux densities per measured magnetizing force. Ferromagnetic materials can also constrain flux along a defined path. Examples of ferromagnetic materials include iron, iron alloys, cobalt, aluminum, and nickel. Many of the other metals can also be classified as ferromagnetics.

Ferromagnetism can't be established solely on the properties of atoms. Instead, the crystalline structure of the metal or alloy is the prime determinant in the formation of ferromagnetism. The crystalline structure or, more specifically, the structure of the atoms within the crystal lattice, is able to alter the spin orientations in such a way as to make certain metals nonferromagnetic (chromium) and others extremely ferromagnetic (iron). A simple expression of this relationship is that a material with a resultant magnetic moment is ferromagnetic, and a material with a vector sum of zero is nonferromagnetic. *Note:* In this example, the vector sum is determined by the magnetic moments.

The potential energy of the crystal lattice also influences the atom orientation along with the subsequent ferromagnetism specification. These orientations are determined on the basis of which state, either magnetic or nonmagnetic, possesses the lowest potential energy. There are two potential energy conditions used for labeling the most stable state. A positive potential energy difference makes the magnetic state the most stable due to the state's lower potential energy. Conversely, in a negative potential energy difference, the unmagnetized state is the most stable. Examples of the influence of this potential energy difference is the ferromagnetic iron and the nonferromagnetic manganese.

The final, uppermost subdivision of matter is the domain. A domain is an independent region within the crystalline structure. Domains of an unmagnetized crystal can still display an individual magnetism. The overall magnetism of the crystal remains unchanged, however, due to the sum of the magnetic moments of the domains equaling zero. An important correlation from this independence is that any material that contains magnetized domains is considered to be a ferromagnetic material.

A domain's crystal lattice is expressed as a clearly defined atomic crystal structure. Some of the more common structural formations for crystals are illustrated in Chapter 2. A body-centered crystal lattice is an example of a structure that is found in the ferromagnetic metal iron. When dealing with crystal domains, there are three axes to study in applying a magnetizing field. These three axes have different degree representations for various crystal lattice structures. In the example of the body-centered iron lattice, the three axes have the values 111, 110, and 100. Likewise, an iron domain's magnetic moments will follow one of these axes based on the direction of the applied magnetic field. In other words, the magnetic moments of the domains will shift to the direction of field application.

Shifting the domain's magnetic moments to fit the direction of the field application falls under three general categories:

+ Shift to the face edge axis = Direction of easy magnetization
+ Shift to the face diagonal axis = Direction of medium magnetization
+ Shift to the cube diagonal axis = Direction of hard magnetization

In the iron example, these directions equal the following axis values:

+ Direction of easy magnetization = 100 axis
+ Direction of medium magnetization = 110 axis
+ Direction of hard magnetization = 111 axis

CRYSTALLINE MATERIALS

Supplementing the basic crystalline structure of ferromagnetic materials like iron are the polycrystalline metals and alloys. These polycrystallines feature nonaligned axis crystals; all of the domains have their magnetic moments aligned along the 100 axis. From this easy magnetization formation, each crystal within this metal is unmagnetized with a random alignment throughout the metal. Thus, there is a similar random magnetic moment distribution in the polycrystalline.

When an external magnetic field is applied to the polycrystalline, there are three effects that determine the final magnetization of the metal. Initially, the domains undergo either an enlargement or a reduction with relation to their individual magnetic moment alignment. This effect makes a modest contribution to the final magnetization of the metal.

The second contributing effect to magnetization is the *Barkhausen effect*. An explanation of the Barkhausen effect relates to a series of incremental "jumps" in the flux density. These stepping flux density jumps correspond to a realignment in the electron spins of a domain. In this effect, there is a simultaneous collective shift in the axis of spin for all of a domain's electrons. Each of these collective axis shifts results in a surge in the crystal's flux density.

The last magnetization effect is an extension of the Barkhausen effect. As the metal reaches its magnetization saturation point, there is a dramatic switch from the jumps in flux density to a smooth transition of all domain magnetic moments. This transition is from being aligned with the crystal's edge to being aligned with the applied magnetic field's direction. While this transition is closely related to the Barkhausen effect, the frequency of the flux density jumps is diminished significantly. The result from this final magnetization effect is the complete magnetization of the target metal.

The relative magnetism in the three basic magnetic materials can be summarized as:

+ Diamagnetic materials—respond weakly to magnetization and exhibit a relative permeability less than unity.
+ Paramagnetic materials—lose complete magnetism in response to an applied external field and have a positive relative permeability.
+ Ferromagnetic materials—retain a degree of magnetism in response to an applied external field, have a strong relative permeability and a significant saturation effect.

One final phenomenon of magnetization deals with the physical alteration of the metal that results from magnetic strains on the crystal. These strains are formed through the realignment of electron spins. As these spins are altered, the balance between the crystal's magnetic and electrical forces is upset. This effect is known as *magnetostriction*. The end result of magnetostriction varies for different metals. Relating to these variations in reaction, there are positive magnetostrictions (a value increase) and negative magnetostrictions (a value decrease). There is a direct relationship between the limitations of magnetostriction and the magnetic saturation of the metal.

The following properties are attributable to ferromagnetic materials:

+ easily magnetized
+ retain magnetization after the removal of the applied field
+ large intrinsic flux density maximum value
+ difficult to demagnetize once they have been magnetized

FREE ENERGY OF A MAGNETIC BODY

The thermodynamics of a magnetic material can be expressed through a series of free energy equations. Beginning with the energy required to raise the field (H_a) and magnetic moments (M) of a magnetic material (V) at a constant temperature:

$$H_a + dH_a = \text{Field increase}$$
$$dM = \text{Magnetic moment change}$$
$$d\,W_{TOT} = V(d(\tfrac{1}{2}\mu_0 H^2) + \mu_0 H_a dM$$

There are two significant terms in this equation. The first term is the work done for increasing the applied field's strength. The second term is the supplied energy for increasing the magnetic moment of the body. This term can also be expressed as dW_m. There is an important relationship between this magnetization work and the typical thermodynamic expression for the work done by an external pressure. Likewise, this relationship can be substituted in the Gibbs free energy equation:

$$G = U - TS + pV$$

as,

$$G = U - TS + pV - \mu_0 H_a M$$

where,

$$U = \text{internal energy}$$
$$S = \text{internal entropy}$$

The changes in G conditions are shown by:

$$dG = dU - TdS - SdT + pdV + Vdp - \mu_0 H_a dM - \mu_0 M dH_a$$

Keeping the temperature and pressure constant while changing the applied field and the magnetic moment:

$$dG = dU - TdS + pdV - \mu_0 H_a dM - \mu_0 M dH_a$$

For a magnetic body, this same relationship is:

$$dU = TdS - pdV + \mu_0 H_a dM$$

In conclusion, the change in the free energy of a body when magnetized to M by an applied field H_a is:

$$G(H) - G(0) = -\mu_0 \int_0^{H_a} M dH_a$$

Appendix C

Supplies Source Guide

References to a number of unusual materials for constructing superconductors have been made in this book. Because some of these materials might be difficult to find in many remote areas, this appendix provides a list of mail-order houses through which these items can be purchased. Additionally, the names and addresses of specific product manufacturers are included.

✤ Carolina Biological Supply Company

East of the Rockies:
2700 York Road
Burlington, NC 27215
(800) 334-5551

West of the Rockies:
Powell Laboratories Division
Box 187
Gladstone, OR 97027
(800) 547-1733

No superconductor laboratory is complete without a set of glassware, several magnets, and a professional-grade scale. Carolina Biological Supply Company offers a complete line of scientific apparatuses coupled with a fast delivery service. A sampling from their current catalog includes:

300 ml Berzelius Beaker, tall form ($2.30)
37x25 mm Stender Dish ($7.45)

150 ml Crucible, low form ($5.00)
145 ml Mortar and Pestle, footed ($16.20)
38 mm Ceramic Magnets ($2.10 per pair)
Triple-Beam Balance ($86.85)

Carolina Biological Supply Company also has a line of A.C.S.-grade chemicals:

500 g Barium Carbonate ($60.00)

✤ Colorado Superconductor, Inc.
P.O. Box 8223
Ft. Collins, CO 80526
(303) 491-9106

An excellent source for domestic supplies of superconductors and superconductor kits. Their current price list includes:

Complete Demonstration Kit ($29)—one 1-inch superconductor, one rare earth magnet, one pair of plastic tweezers, and complete documentation.
Advanced Demonstration Kit ($69)—all of the above, plus a critical temperature determination thermocouple.
Lab Kit for Six Groups ($99)—the basic kit for six experimenters.
Advanced Lab for Six Groups ($329)—the advanced kit for six experimenters.
1-inch superconductor pellets (2 for $35)
Ceramic Source Material ($1 per gram)

✤ Furuuchi Chemical Corporation
2-7-12, Omori-kita
Ota-ku, Tokyo   JAPAN 143
(03) 762-8161

An elaborate series of kits dealing with demonstrations of the Meissner effect. Their current catalog lists:

UFO FINE-90 Set A (13,000 yen)—one superconductor, five magnets, one set of lab dishes, and one pair of plastic tweezers.
UFO FINE-90 Kit Set B (30,000 yen)—all the materials of Set A plus one metal dish, one pair of leather gloves, and one pair of goggles.
UFO FINE-90 Kit Set C (95,000 yen)—all of the materials from Set A and Set B, plus one 5-liter metal vacuum jar.

5-liter vacuum jar (68,000 yen)
10-liter vacuum jar (84,000 yen)
20-liter vacuum jar (116,000 yen)

✢ Edmund Scientific Co.
101 E. Gloucester Pike
Barrington, NJ 08007
(609) 573-6250

An entire line of alnico and rare earth magnets. A sample of the diverse selection offered by Edmund Scientific includes:

Circular Ceramic Barium Iron Oxide Magnet ($27.95)
Circular 1-inch Diameter Ferrite Magnet ($5.50 per twenty)
Disc 1.5-inch Diameter Alnico Magnet ($5.00 per four)
Disc 1-inch Diameter Cobalt Magnet ($41.95)
Disc 1-inch Diameter Neodymium-Iron-Boron Magnet ($60.00)

✢ Linear Research, Inc.
5231 Cushman Place, Suite 21
San Diego, CA 92110
(619) 299-0719

Manufacturer of the LR-400 Four-Wire AC Resistance & Mutual Inductance Bridge for non-contact transformer measurements. Also, the LR-4PC is an accessory for interfacing an IBM PC.

Appendix D

Time Line of Significant Events in Superconductor Research

The following chronology represents important events in the history of superconductor research.

1908 Kammerlingh Onnes liquifies helium.

1911 Kammerlingh Onnes reports a "superconductive" state for mercury at 4.2 °K.

1933 Meissner and Ochsenfeld discover the Meissner effect. The critical temperature for superconductivity is raised to 10 °K.

1941 The critical temperature is elevated to 15 °K.

1950 Maxwell verifies the existence of the isotope effect. Fröhlich postulates an electron-phonon interaction in superconductors.

1953 The critical temperature is increased to 17 °K.

1955 Schafroth offers a definition of the perfect superconductor.

1956 Cooper theorizes the presence of Cooper pairs in superconductor crystal lattices.

1957 Bardeen, Cooper, and Schrieffer develop their theory of superconductivity.

1960 The critical temperature is raised to 18 °K. Buchhold designs a superconductor-based magnetic bearing.

1973 The critical temperature is increased to 23 °K.

1986 January: Müller and Bednorz discover a superconductive ceramic material with a critical temperature of 30 °K. December: The critical temperature is elevated to 39 °K.

1987 February: Chu modifies the Müller and Bednorz ceramic for a heightened critical temperature of 98 °K. June: The critical temperature is experimentally tested at 200 °K.

1988 March: University of Arkansas-Fayetteville develops a ceramic with a critical temperature of 106 °K. March: The IBM Corp. Almaden Research Center creates a thalium, barium, calcium, and copper oxide ceramic with a critical temperature of 125 °K.

Appendix E

Major Players in Superconductivity Research

A number of different companies are currently conducting extensive research in superconductivity. The following list is only a portion of the complete superconductivity research picture.

AT&T BELL LABORATORIES
Short Hills, NJ

AMERICAN SUPERCONDUCTOR CORPORATION
Boston, MA

ARGONNE NATIONAL LABORATORY
Argonne, IL

CERACON, INC.
Sacramento, CA

CONDUCTIS
Palo Alto, CA

GA TECHNOLOGIES
San Diego, CA

GENERAL DYNAMICS
San Diego, CA

GENERAL ELECTRIC MEDICAL SYSTEMS GROUP
Milwaukee, WI

GENERAL ELECTRIC RESERACH AND DEVELOPMENT CENTER
Schenectady, NY

GENERAL MOTORS RESEARCH LABORATORIES
Warren, MI

GEORGIA INSTITUTE OF TECHNOLOGY
Atlanta, GA

HITACHI, LTD.
Japan

HYPRES, INC.
Elmsford, NY

IBM CORP. ALAMADEN RESEARCH CENTER
San Jose, CA

IBM RESEARCH DIVISION
Yorktown Heights, NY

INTERMAGNETICS GENERAL CORPORATION
Guilderland, NY

MASSACHUSETTS INSTITUTE OF TECHNOLOGY
Cambridge, MA

MAX PLANCK INSTITUTE
Stuttgart, West Germany

MICROELECTRONICS AND COMPUTER TECHNOLOGY CORP.
Austin, TX

MONOLITHIC SUPERCONDUCTORS, INC.
Beaverton, OR

NATIONAL BUREAU OF STANDARDS
Boulder, CO

SUPERCON, INC.
Shrewsbury, MA

Superconductor Technology, Inc.
Santa Barbara, CA

TRW, Inc., Space and Technology Group
Redondo Beach, CA

Westinghouse
Pittsburgh, PA

For Further Reading

BOOKS

Introduction to Ferromagnetism, 1927, BITTER, F.
  MCGRAW-HILL BOOK COMPANY, INC., NEW YORK, NY
  ✢ Many of the important early theories of magnetism and magnet preparation methodology are outlined in this text.

Theory of Superconductivity, 1964, BLATT, J.M.
  ACADEMIC PRESS, NEW YORK, NY
  ✢ A short study of the superconductivity based largely on early research.

The Theory of Superconductivity, 1962, BOGOLIUBOV, N.N. (EDITOR)
  GORDON & BREACH, NEW YORK, NY
  ✢ A collection of papers exploring research into superconductivity.

Design and Fabrication of Conventional and Unconventional Superconductors, 1984, COLLINGS, E.W.
  NOYES PUBLICATIONS, PARK RIDGE
  ✢ Many of the considerations that enter into the construction of superconductors are discussed in this book.

Superconductivity, 1972, FIRTH, I.M.
  M&B MONOGRAPHS, MILLS & BOON LIMITED, LONDON, ENGLAND
  ✢ A thin account of superconductivity research from a distinctly British point of view.

A Guide to Superconductivity, 1969, FISHLOCK, D. (EDITOR)
MACDONALD & CO. LTD., LONDON, ENGLAND
  ✛ A series of invited papers covering all aspects of superconductivity. This is a good introductory text.

Behavior of Electrons in Atoms, 1962, HOCHSTRASSER, R. M.
W. A. BENJAMIN, INC., NEW YORK, NY
  ✛ Provides a solid understanding of quantum mechanics.

Atomic Structure, 1938, LOEB, L. B.
JOHN WILEY & SONS, NEW YORK, NY
  ✛ This book offers a superb introduction to the Bohr atom and wave-mechanic mathematical concepts.

Superconductivity: The Threshold of a New Technology, 1988, MAYO, J.
TAB BOOKS, INC., CATALOG #3022
  ✛ This is an easy-to-read nontechnical look at superconductivity and its associated research.

Matter at Low Temperatures, 1984, MCCLINTOCK, P.V.E., ET AL.
BLACKIE & SON, LONDON, ENGLAND
  ✛ The exact nature of matter on the atomic level at low temperatures is the subject of this text.

Introduction to Superconductivity, 1978, ROSE-INNES, A.C. AND E.H. RHODERICK
PERGAMON PRESS, OXFORD, ENGLAND
  ✛ A highly technical look into the thermodynamics and the microscopic theory of superconductivity.

Type II Superconductivity, 1969, SAINT-JAMES, D., G. SARMA, AND E.J. THOMAS
PERGAMON PRESS, OXFORD, ENGLAND
  ✛ This is a detailed technical concentration on one group of superconductors.

Principles of Superconductive Devices and Circuits, 1981, VAN DUZER, T. AND C.W. TURNER
ARNOLD PUBLISHERS, LONDON, ENGLAND
  ✛ Many of the higher-temperature superconductors are the subject of this book.

Basic Chemical Thermodynamics, WASER, J.
W. A. BENJAMIN, INC., NEW YORK, NY
  ✛ An excellent introductory text.

Superconductivity and Its Applications, 1970, WILLIAMS, J.E.C.
  PION LIMITED, LONDON, ENGLAND
  ✤ How practical have applications with superconductors been?
    This book examines all of the major uses for superconductivity.

SUPPLEMENTAL REFERENCES ———————————————————————

*Superconductivity in D- & F-Band Metals: Proceedings of the AIP
  Conference, Univ. of Rochester, 1971*, 1972, AMERICAN INSTITUTE
  OF PHYSICS, (375 pp., $14.00)
AMERICAN INSTITUTE OF PHYSICS
1. SUPERCONDUCTIVITY. 2. ENERGY-BAND THEORY OF SOLIDS. I. Douglass, D. H.,
editor. II. Title
ISBN 0-88318-103-7

Theory of Free Radical Polymerization, 1968, BAGDASAR'YAN, K. S. (328
  pp., $64.00)
CORONET BOOKS
1. SUPERCONDUCTIVITY. 2. POLYMERS AND POLYMERIZATION. I. Title.
ISBN 0-7065-0604-9 Hardcover text edition.

Superconductivity, 1987, BUSINESS COMMUNICATIONS STAFF   (paper:
  $1,950.00)
BUSINESS COMMUNICATIONS COMPANY, INCORPORATED
1. SUPERCONDUCTIVITY. I. Title.
ISBN 0-89336-616-1 Paper.

The Superconductivity Industry, 1980, BUSINESS COMMUNICATIONS
  STAFF (124 pp., $750.00)
BUSINESS COMMUNICATIONS COMPANY, INCORPORATED, 1980.
1. SUPERCONDUCTIVITY. I. Title
ISBN 0-89336-144-5

Superconductivity in Science & Technology, COHEN, M. H. (EDITOR) (paper: $42.80)
BOOKS ON DEMAND
1. SUPERCONDUCTIVITY. I. Title.
ISBN 0-317-08095-4   Paper.

Advances in Superconductivity, 1983, DEAVER, B. (EDITOR) (538 pp.,
  $85.00)
PLENUM PUBLISHING CORPORATION
(NATO ASI SERIES B, PHYSICS; VOL. 100)
1. SUPERCONDUCTIVITY. I. Ruvalds, John, editor. II. Title. III. Series.
ISBN 0-306-41388-4

Future Trends in Superconductive Electronics, 1979, DEAVER, B. S., JR.
  (EDITOR) (Library binding—adult: $22.00)

AMERICAN INSTITUTE OF PHYSICS
ill. (AIP Conference Proceedings; No. 44)
1. SUPERCONDUCTIVITY. I. Falco, C. M., editor. II. Harris, J. H., editor. III. Wolf, S.
A., editor. IV. Title. V. Series.
ISBN 0-88318-143-6

*Superconductivity in Ternary Compounds I: Structural, Electronics &
    Lattices Properties*, 1982, FISCHER, O. (EDITOR)   (320 pp., $43.50)
SPRINGER-VERLAG NEW YORK, INCORPORATED
ill.—(Topics in Current Physics Ser.; Vol. 32)
1. SUPERCONDUCTIVITY. I. Maple, M. B., editor. II. Title. III. Series.
ISBN 0-387-11670-2

Superconductivity & Quantum Fluids, 1970, GALASIEWICZ, Z. M.
    ($34.00)
PERGAMON BOOKS, INCORPORATED
1. SUPERCONDUCTIVITY. 2. QUANTUM LIQUIDS. I. Title.
ISBN 0-08-013089-5

Nonequilibrium Superconductivity, 1987, GINZBURG, V. L., (EDITOR)
    (282 pp., Hardcover $89.00)
NOVA SCIENCE PUBLISHERS, INCORPORATED
(PROCEEDINGS OF THE LEBEDEV PHYSICS INSTITUTE OF THE ACADEMY OF
SCIENCES OF THE U.S.S.R. SER.; VOL. 174)
1. SOLID STATE PHYSICS. 2. SUPERCONDUCTIVITY. I. Title. II. Series.
ISBN 0-941743-09-8

High-Temperature Superconductivity, 1982, GINZBURG, V. L. (EDITOR)
    (364 pp., $59.50)
PLENUM PUBLISHING CORPORATION
1. MATERIALS AT HIGH TEMPERATURES. 2. SUPERCONDUCTIVITY. I. Kirzhnits,
D. A., editor. II. Agyei, A. K., translator. III. Title.
ISBN 0-306-10970-0

Percolation, Localization, & Superconductivity, 1984, GOLDMAN, A. M.
    (EDITOR) (472 pp., $85.00)
PLENUM PUBLISHING CORPORATION
(NATO ASI SERIES B, PHYSICS; VOL. 109)
1. SUPERCONDUCTIVITY. I. Wolf, Stuart A., editor. II. Title. III. Series.
ISBN 0-306-41713-8

Nonequilibrium Superconductivity, Phonons, & Kapitza Boundaries,
    1981, GRAY, K. E. (EDITOR) (710 pp., $110.00)
PLENUM PUBLISHING CORPORATION, 06/1981.
(NATO ASI SERIES B, PHYSICS; VOL. 65)
1. SUPERCONDUCTIVITY. 2. PHONONS. I. Title. II. Series.
ISBN 0-306-40720-5

SQUID-Superconducting Quantum Interference Devices & Their Applications: International Conference, Oct. 4-8, 1976, Berlin, 1977, HAHLBOHM, H. D. (EDITOR) ($92.00)
WALTER, DE GRUYTER, INCORPORATED
1. SUPERCONDUCTIVITY. I. Lubbing, H., editor. II. Title.
ISBN 3-11-006878-8

Nonequilibrium Superconductivity, 1986, LANGENBERG, D. N. (EDITOR) (710 pp., $120.50)
ELSEVIER SCIENCE PUBLISHING COMPANY, INCORPORATED
(MODERN PROBLEMS IN CONDENSED MATTER SCIENCES SER.; VOL. 12)
1. SUPERCONDUCTIVITY. 2. SUPERCONDUCTORS. I. Larken, A. I., editor. II. Title.
III. Series.
ISBN 0-317-45880-9

Method of Correlation Function in Superconductivity Theory, 1971, LUEDERS, G. ($56.70)
SPRINGER-VERLAG NEW YORK, INCORPORATED
ill. (Springer Tracts in Modern Physics Vol. 56)
1. SUPERCONDUCTIVITY. I. Hoehler, G., editor. II. Title. III. Series.
ISBN 0-387-05251-8

Superconductivity in Ternary Compounds II: Superconductivity & Magnetism, 1982, MAPLE, M. B. (EDITOR) (335 pp., $43.50)
SPRINGER-VERLAG NEW YORK, INCORPORATED
ill.—(Topics in Current Physics; Vol. 34)
1. SUPERCONDUCTIVITY. 2. NONFERROUS METALS. I. Fischer, O., editor. II. Title.
III. Series.
ISBN 0-387-11814-4

Superconducting in Magnetic & Exotic Materials: Proceedings of the Sixth Taniguchi International Symposium, Kashikojima, Japan, Nov. 14-18, 1983, 1984, MATSUBARA, T. (EDITOR) (225 pp., $32.50)
SPRINGER-VERLAG NEW YORK, INCORPORATED
ill.—(Springer Series in Solid-State Sciences Ser.; Vol. 52)
1. SUPERCONDUCTIVITY. 2. MAGNETIC MATERIALS. I. Kotani, A., editor. II. Title. III.
Series.
ISBN 0-387-13324-0

Superconductivity & Superconducting Materials, 1983, NARLIKAR, A. V. (306 pp., $39.00)
ENGINEERING PUBLICATIONS
ill.—(Solid Physics Ser.; No. 1)
1. SUPERCONDUCTIVITY. I. Ekbote, S. N. II. Title. III. Series.
ISBN 0-9605004-9-9

Applied Superconductivity, 1975, New House, V.
Academic Press, Incorporated
1. SUPERCONDUCTIVITY. I. Title.
ISBN 0-12-517701-1: $98.00. ISBN 0-12-517702-X: $85.50.

Superconductivity, Parks, R. D. (Editor) (Paper: $160.00)
Books on Demand
ill.
1. SUPERCONDUCTIVITY. I. Title.
ISBN 0-317-07985-9

Superconductivity Parks, R. D. (Editor) (Paper: $160.00)
Books on Demand
1. SUPERCONDUCTIVITY. I. Title
ISBN 0-317-08358-9   Paper.

Superconductivity—2nd ed., 1960-1965, Shoenberg, D. (Paper: $7.95)
Cambridge University Press
ill. (Cambridge Monographs on Physics)
1. ELECTROMAGNETIC THEORY. 2. SUPERCONDUCTIVITY. 3. LOW TEMPERATURE
RESEARCH. I. Title. II. Series.
ISBN 0-521-09254-X

Superconductivity—2nd ed., Shoenberg, D. (Paper: $67.50)
Books on Demand
(Cambridge Monographs on Physics)
1. SUPERCONDUCTIVITY. I. Title. II. Series.
ISBN 0-317-09142-5

Superconductivity in D- & F-Band Metals, 1980, Suhl, H. (Editor)
($65.50)
Academic Press, Incorporated
1. SUPERCONDUCTIVITY. I. Maple, M. Brian, Editor, II. Title.
ISBN 0-12-676150-7

Superconductors: Proceedings, Tanenbaum, M. (Editor) (Paper:
$40.30)
Books on Demand
1. SUPERCONDUCTIVITY. I. Wright, W. V., editor. II. Title.
ISBN 0-317-08032-6

Superconductivity, 1970, Taylor, A. W. (110 pp., $9.95)
Taylor & Francis, Incorporated
(Wykeham Science Ser.; No. 11)
1. SUPERCONDUCTIVITY. I. Noakes, G. R. II. Title. III. Series.
ISBN 0-8448-1113-0

Superconductivity, 1970, TAYLOR, A. W. (110 pp., Paper: $16.00)
TAYLOR & FRANCIS, INCORPORATED
1. SUPERCONDUCTIVITY. I. Title.
ISBN 0-85109-120-2

Superfluidity & Superconductivity—2nd ed., 1986, TILLEY, D. R.
    (440 pp.)
TAYLOR & FRANCIS, INCORPORATED
ill.—(Graduate Student Series in Physics)
1. SUPERFLUIDITY. 2. SUPERCONDUCTIVITY. I. Tilley, J. II. Title. III. Series.
ISBN 0-85274-807-8: $77.00. ISBN 0-85274-791-9   Paper: $40.00.

Introduction to Superconductivity, 1980, TINKHAM, M., KRIEGER,
    ROBERT E. (312 pp., Library binding—adult: $24.50)
PUBLISHING COMPANY, INCORPORATED
1. SUPERCONDUCTIVITY. I. Bayne, Bradford, editor. II. Gardner, Michael, editor. III. Title.
ISBN 0-89874-049-5

Superconductivity, 1964, TINKHAM, M. (142 pp., Paper: $32.00)
GORDON & BREACH SCIENCE PUBLISHERS, INCORPORATED
(Documents on Modern Physics Ser.)
1. SUPERCONDUCTIVITY. I. Title. II. Series.
ISBN 0-677-00065-0

Superconductivity of Transition Metals: Their Alloys & Compounds,
    1982, VONSOVSKY, S. V. (512 pp., $47.20)
SPRINGER-VERLAG NEW YORK, INCORPORATED
ill.—(Springer Series in Solid-State Sciences; Vol. 27)
1. SUPERCONDUCTIVITY. 2. TRANSITION METALS. I. Title. II. Series.
ISBN 0-387-11382-7

Superconductivity: McGill Summer School Proceedings, 1969,
    WALLACE, P. R. (EDITOR)
GORDON & BREACH SCIENCE PUBLISHERS, INCORPORATED
1. SUPERCONDUCTIVITY. I. TITLE.
ISBN 0-677-13810-5: $160.00. ISBN 0-677-13820-2: $122.00.
ISBN 0-677-13210-7: $250.00.

MAGAZINE ARTICLES

ANDERSON AND ABRAHAMS, "Superconductivity Theories Narrow
    Down," *Nature*, 4 Jun 1987, p. 363.

BARDEEN, COOPER, AND SCHRIEFFER, *Phys. Rev.*, 108 (1957), p. 1175.

BRANDT AND GINSBURG, "Superconductivity at High Pressure," *Sci. Am.*,
    April 1971, p. 83.

COHEN, "Big Steps in HEMTs, Josephson Junctions," *Electronics*, 3 March 1988, p. 34.

COOPER, *Phys. Rev.*, 104 (1956), p. 1189.

ESSMANN AND TRAUBLE, "The Magnetic Structure of Superconductors," *Sci. Am.*, March 1971, p. 75.

GREGORY, "Applications and Fabrication Processes of Superconducting Composite Materials," *Journal of Metals*, Jun 1984, p. 30.

HARTLEY, "High-Temperature Superconductivity: What's Here, What's Near, and What's Unclear," *Science News*, 15 Aug 1987, p. 106.

LINEBACK AND NAEGELE, "Superconductor Race Attracts MCC," *Electronics*, 25 Jun 1987, p. 33.

LOUNASMAA, "New Methods for Approaching Absolute Zero," *Sci. Am.*, Dec 1969, p. 26.

MAYBURY, "The Language of Quantum Mechanics," *J. Chem. Educ.*, 39 (1962), p. 367.

NASH, "Elementary Thermodynamics," *J. Chem. Educ.*, 42 (1965), p. 64.

OLSEN, "Crystal Models," *J. Chem. Educ.*, 44 (1967), p. 728.

SCHECHTER, "How to Make Your Own Superconductors," *Omni*, Nov 1987, p. 72.

SHOCKLEY, "The Quantum Physics of Solids-I," *Bell System Technical Journal*, 18 (1939), p. 645.

THOMSEN, "Current News About Superconductors," *Science News*, 16 May 1987, p. 308.

WALLER, "Will 1988 See a 92° K Superconductor IC?," *Electronics*, 7 Jan 1988, p. 32.

ADDITIONAL SOURCES OF INFORMATION ————————————

Physics Today

This journal contains extensive coverage of superconductivity on a monthly basis.

Superconductors Update

A bimonthly abstract summary of current research in superconductivity. You can order a copy from:

STN International Marketing
Dept. 35787
2540 Olentangy River Road
P.O. Box 02228
Columbus, OH 43202

Glossary

absolute zero—the temperature at which a material has no molecular movement and generates no heat; $0°K = -273.15°C$.

access time—the delay time interval between the loading of a memory location and the latching of the stored data.

address—the location in memory where a given binary bit or word of information is stored.

allophone—two or more variants of the same phoneme.

alphanumeric—the set of alphabetic, numeric, and punctuation characters used for computer input.

analog/digital (A/D) conversion—a device that measures incoming voltages and outputs a corresponding digital number for each voltage.

ASCII—American Standard Code for Information Interchange.

assembly language—A low-level symbolic programming language that comes close to programming a computer in its internal machine language.

Bardeen, Cooper, and Schrieffer Theory—the 1957 theory of superconductivity formed by John Bardeen, Leon Cooper, and J. R. Schrieffer.

binary—the base-two number system in which 1 and 0 represent the ON and OFF states of a circuit.

bit—one binary digit.

byte—a group of eight bits.

CCD—charge-coupled device; a SAM with slow access times.

chip—an integrated circuit.

chip enable—a pin for activating the operations of a chip.

chip select—a pin for selecting the I/O ports of a chip.

conductor—a material that carries electricity.

Cooper pairs—electron pairs in superconductors.

CMOS—a complementary *metal-oxide semiconductor* IC that contains both p-channel and n-channel MOS transistors.

CPU—central *processing unit*; the major operations center of the computer where decisions and calculations are made.

critical field—the maximum magnetic field strength that perpetuates superconductivity.

critical temperature—the maximum temperature at which superconductivity occurs in a substance.

current density—the maximum current that can be carried by a superconductor.

data—information that a computer manipulates.

data rate—the amount of data transmitted through a communications line per unit of time.

debug—to remove program errors, or bugs, from a program.

digital—a circuit that has only two states, ON and OFF, which are usually represented by the binary number system.

disk—the magnetic media on which computer programs and data are stored.

DOS—*disk operating system*; allows the use of general commands to manipulate the data stored on a disk.

EAROM—electrically alterable *read-only memory*; also known as *read mostly memory.*

EEPROM—electrically erasable *programmable read-only memory*; both read and write operations can be executed in the host circuit.

electron—a negatively-charged atomic particle.

EPROM—an erasable *programmable read-only memory* semiconductor that can be user-programmed.

field-programmable logic array—a logical combination of programmable AND/OR gates.

firmware—software instructions permanently stored within a computer using a read-only memory (ROM) device.

floppy disk—see *disk.*

flowchart—a diagram of the various steps to be taken by a computer in running a program.

hardware—the computer itself and its associated peripherals, as opposed to the software programs that the computer uses.

hexadecimal—a base-16 number system often used in programming assembly language.

input—to send data into a computer.

input/output (I/O) devices—peripheral hardware devices that exchange information with a computer.

insulator—a material that does not conduct electricity.

interface—a device that converts electronic signals to enable communications between two devices; also called a *port*.

Josephson junction—two superconductors separated by a thin insulating barrier that acts as an electronic switch.

languages—the set of words and commands that are understood by the computer and used in writing a program.

lattice—a three-dimensional atomic arrangement in a solid.

loop—a programming technique that allows a portion of a program to be repeated several times.

LSI—*large-scale integration*; a layered semiconductor fabricated from approximately 10,000 discrete devices.

machine language—the internal, low level language of the computer.

memory—an area within a computer reserved for storing data and programs the computer can operate on.

microcomputer—a small computer, such as the IBM PC AT, that contains all of the instructions it needs to operate on a few internal integrated circuits.

mnemonic—an abbreviation or word that represents another word or phrase.

MOS—a *metal-oxide semiconductor* containing field-effect MOS transistors.

MRI—*magnetic resonance imaging*.

meissner effect—the expulsion of a magnetic field from a superconductor.

NMOS—an *n*-channel *metal-oxide semiconductor* with n-type source and drain diffusions in a p substrate.

nonvolatile—the ability of a memory to retain its data without a power source.

octal—a base-eight number system often used in machine language programming.

opcode—an operation code signifying a particular task to be performed by the computer.

parallel port—a data communications channel that sends data out along several wires so that entire bytes can be transmitted simultaneously, rather than by a single bit at a time.

peripheral—an external device that communicates with a computer, such as a printer, a modem, or a disk drive.

phonon—the atomic interactive device which initiates electron pairing in a superconductor.

PLA—see *field-programmable logic array*.

PMOS—a *p*-channel *metal oxide semiconductor* with p-type source and drain diffusions in an n substrate.

program—a set of instructions for a computer to perform.

RAM—*random access memory*; integrated circuits within the computer where data and programs can be stored and recalled. Data stored within RAM is lost when the computer's power is turned off.

ROM—*read-only memory*; integrated circuits that permanently store data or programs. The information contained on a ROM chip cannot be changed and is not lost when the computer's power is turned off.

RS-232C—a standard form for serial computer interfaces.

serial communications—a method of data communication in which bits of information are sent consecutively through one wire.

software—a set of programmed instructions that the computer must execute.

SQUID—*superconducting quantum interference device*.

statement—a single computer instruction.

static—a RAM whose data is retained over time without the need for refreshing.

subroutine—a small program routine contained within a larger program.

terminal—an input/output device that uses a keyboard and a video display.

Type I superconductor—superconductors formed from simple metals with one critical field. Type I superconductors emit a magnetic field until the critical field is reached.

Type II superconductor—superconductors formed from oxides and metals with two critical fields. Type II superconductors maintain partial superconducting properties while exposed to a magnetic field that is stronger than the superconductor's critical field.

volatile—the inability of a memory to retain its data without a power source.

word—a basic unit of computer memory usually expressed in terms of a byte.

Index

Notes

Notes

Other Bestsellers from TAB

☐ **EXPERIMENTS IN ARTIFICIAL NEURAL NETWORKS**—*Ed Rietman*

Build your own neural networking breadboards—systems that can store and retrieve information like the brain! This book shows you how to use threshold logic circuits and computer software programs to simulate the neural systems of the brain in information processing. The author describes artificial electronic neural networks and provides detailed schematics for the construction of six neural network circuits. The circuits are stand-alone and PC-interfaced units.160 pp., 55 illus.

Paper $19.95   **Hard $24.95**
Book No. 3037

☐ **FIBEROPTICS AND LASER HANDBOOK**—2nd Ed.—*Edward L. Safford, Jr. and John A. McCann*

Explore the dramatic impact that lasers and fiberoptics have on our daily lives—PLUS, exciting ideas for your own experiments! Now, with the help of experts Safford and McCann, you'll discover the most current concepts, practices, and applications of fiberoptics, lasers, and electromagnetic radiation technology. Included are terms and definitions, discussions of the types and operations of current systems, and amazingly simple experiments you can conduct! 240 pp., 108 illus.

Paper $19.95   **Hard $24.95**
Book No. 2981

☐ **LASERS—THE LIGHT FANTASTIC**—2nd Edition—*Clayton L. Hallmark and Delton T. Horn*

Gain insight into all the various ways lasers are used today . . . in communications, in radar, as gyroscopes, in industry, and in commerce. Plus, more emphasis is placed on laser applications for electronics hobbyists and general science enthusiasts. If you want to experiment with lasers, you will find the guidance you need here—including safety techniques, a complete glossary of technical terms, actual schematics, and information on obtaining the necessary materials. 280 pp., 129 illus.

Paper $15.95   **Hard $19.95**
Book No. 2905

☐ **SUPERCONDUCTIVITY—THE THRESHOLD OF ANEW TECHNOLOGY**—*Jonathan L. Mayo*

Superconductivity is generating an excitement in the scientific world not seen for decades! Experts are predicting advances in state-of-the-art technology that will make most existing electrical and electronic technologies obsolete! This book is one of the most complete and thorough introductions to a multifaceted phenomenon that covers the full spectrum of superconductivity and superconductive technology. 160 pp., 58 illus.

Paper $14.95   **Hard $18.95**
Book No. 3022

☐ **EXPERIMENTS WITH EPROMS**—*Dave Prochnow*

One of the greatest versatilities in advanced circuit design is EPROM (Erasable-Programmable Read-Only-Memory) programming. Now, Dave Prochnow takes an in-depth look at these special integrated circuits (ICs) that can be user-programmed to perform specific applications in a microcomputer. Fifteen fascinating experiments are a special feature of this book that presents not only the technology but also explains the use of EPROMs. 208 pp., 241 illus.

Paper $19.95   **Hard $24.95**
Book No. 2962

☐ **UNDERSTANDING MAGNETISM: MAGNETS, ELECTROMAGNETS AND SUPERCONDUCTING MAGNETS**—*Robert Wood*

Explore the mysteries of magnetic and electromagnetic phenomena. This book bridges the information gap between children's books on magnets and the physicists' technical manuals. Written in an easy-to-follow manner, *Understanding Magnetism* examines the world of magnetic phenomena and its relationship to electricity. Thirteen illustrated experiments are provided to give you hands-on understanding of magnetic fields. 176 pp., 138 illus.

Paper $13.95   **Hard $17.95**
Book No. 2772

APR 4 1990

4/00

7\16

Other Bestsellers from TAB

☐ **BUILD YOUR OWN LASER, PHASER, ION RAY GUN AND OTHER WORKING SPACE-AGE PROJECTS**—*Robert E. Iannini*

Here's the highly skilled do-it-yourself guidance that makes it possible for you to build such interesting and useful projects as a burning laser, a high power ruby/YAG, a high-frequency translator, a light beam communications system, a snooper phone listening device, and much more—24 exciting projects in all! 400 pp., 302 illus.

Paper $13.95   **Hard $16.95**
Book No. 1604

☐ **HANDBOOK OF REMOTE CONTROL AND AU-TOMATION TECHNIQUES**—2nd Edition—*John E. Cunningham and Delton T. Horn*

From how-to's for analyzing your control needs to coming up with the electronic and mechanical systems to do the job, the authors provide a wealth of information on just about every kind of remote control system imaginable; temperature, light, and tone sensitive devices; pressure and gas sensors; radio controlled units; time controlled systems; and microcomputer interface systems; even robots. 350 pp., 306 illus.

Paper $14.95   **Hard $18.95**
Book No. 1777

Send $1 for the new TAB Catalog describing over 1300 titles currently in print and receive a coupon worth $1 off on your next purchase from TAB.

(In PA, NY, and ME add applicable sales tax. Orders subject to credit approval. Orders outside U.S. must be prepaid with international money orders in U.S. dollars.)

*Prices subject to change without notice.

━━

To purchase these or any other books from TAB, visit your local bookstore, return this coupon, or call toll-free 1-800-233-1128 (In PA and AK call 1-717-794-2191).

Product No.	Hard or Paper	Title	Quantity	Price

☐ Check or money order enclosed made payable to TAB BOOKS Inc.

Charge my   ☐ VISA   ☐ MasterCard   ☐ American Express

Acct. No. _____ Exp. _____

Signature _____

Please Print
Name _____

Company _____

Address _____

City _____

State _____ Zip _____

Subtotal	
Postage/Handling ($5.00 outside U.S.A. and Canada)	$2.50
In PA add 6% sales tax	
TOTAL	

Mail coupon to:

TAB BOOKS Inc.
Blue Ridge Summit
PA 17294-0840

BC